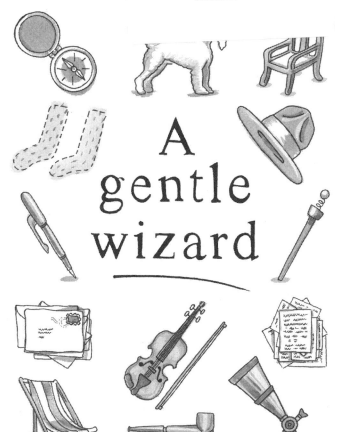

A
gentle
wizard

FIGHTING GRAVITY — AND WINNING!

A gentle wizard

written by
Nils Andersson

illustrated by
Oliver Dean

speed of think
publishing

A GENTLE WIZARD

Published by Speed of Think Publishing Ltd

Text copyright © Nils Andersson 2017
Image copyright © Oliver Dean 2017

ISBN: 978-0-9956462-0-9

PROLOGUE
An early morning walk

THE YOUNG MAN closed the door to the apartment, carefully trying not to make a sound. It had been a rough night and he did not want to disturb his sleeping wife. The baby had kept them up, not settling until the sun crawled across the horizon. By then it had been time for him to get out of bed, anyway. He rubbed his tired eyes to help them adjust to the gloomy darkness of the stairwell. Stifling a yawn he started down the stairs, one step at a time.

The heavy wooden door opened to a different world, a bright bustling world that had long left sleep behind. The air was fresh. Merchants and peddlers were busy setting up their stalls in the arcades. An old tram rattled past. People hurried along the cobbled street on their way to work.

He stepped out into the morning sun. Suddenly he did not feel quite so tired anymore. Drawn in by a mouth-watering smell, he stopped at the baker's to pick up some breakfast. Taking a bite out of a fresh piece of bread, still warm from the oven, he set off down the street.

He looked up at the ornate clock that decorated the medieval tower at the far end of the square. For a moment he was blinded, as the morning sun

1

reflected off the shiny gold on the old clock. *Infernal light*, he thought with a wry smile, *always causing trouble*. After blinking a couple of times, he looked up at the clock again. He was late, but the office could wait.

There was no need to rush. He knew he would get the work done before the end of the day. He enjoyed the job at the patent office and took it seriously, but much of it was mechanical with little thinking required. He liked that. He preferred to save his real thinking for himself.

"Time is an illusion, anyway," he chuckled as he continued down the street.

As he turned the corner by the clock tower, he was reminded how the idea had come to him. How he had realised the old masters had got it all wrong. The mechanical clockwork universe, wound up at some point in the distant past and now ticking away in an orderly fashion. This was not how it worked at all.

It had taken him a long time to figure out the details. Hours and hours of scribbling down ideas on pieces of paper. Late nights struggling with messy mathematics. But he did work it out. The theory was almost complete. A neatly typed manuscript mingled with the blueprints for various crackpot inventions in his briefcase. He grabbed the bag just a little bit tighter. "My work and my passion," he mumbled to himself.

It was such a simple question. What happens if you ride along a beam of light? And the answer turned out to be simple, too. Although perhaps a bit confusing. *I guess it is relative*, he thought as he looked back on the nearly ten years it had taken to figure it out.

Nothing happens when you ride alongside a beam of light, because you can't.

At first he had not been able to decide if he liked this answer. It felt like cheating, like claiming there was no question in the first place. But the more he thought about it, the more sense it made. As soon as you started thinking about light as a wave of electricity and magnetism, rather than a bunch of bullet-like particles, it became obvious. There had to be something bobbing up and down, or maybe sideways, as the wave moved along. If you were to ride along with it, this motion would stop. There would not be a wave anymore. The light would stop and that did not seem right.

The answer made sense but it was problematic.

He recalled the early morning walk when he had tried to figure out what the world would look like to someone that was moving along with the light. If he were to move away from the clock tower at this astounding speed, the clock would appear to stop. As he caught up with the beam of light he would not see the clock advance at all. Light reflected from the hands on the clock would never catch up with him. Time would freeze. He

remembered thinking this must be nonsense. But
then again... you could not argue with the logic.

This was the breakthrough. As soon as he gave
up on the idea that time was the same for everyone, the pieces of the puzzle came together.

Every moving body has its own measure of
time, its own clock, and how it measures length,
breadth, and thickness is different, as well. It really is all relative.

Time is different for someone riding in a speeding automobile, than for someone having a rest in
a cafe. Space is relative, too. Moving objects ap-

4

pear shorter than they are if you hold them down and measure them. Under normal circumstances, the effect is so tiny it makes no difference. But if you get close to the speed of light it becomes significant. And when you reach the speed of light... Well, you can't. Nothing can catch light. There is a limit to how fast things can go.

A cosmic speed limit! He liked the sound of that.

Perhaps the slowing down of clocks is not so strange, after all. It is just a matter of perspective, similar to when two people of the same height walk away from each other, then stop and look back to find that each appears shorter. This change in size does not strike us as odd simply because we are used to it.

He still had to figure out how you come by this individual time and space. This was a tricky, but the pieces fell into place in a way that seemed natural, making him confident that the answer had to be right.

You have to combine space and time so that you measure up and down, left and right, backwards and forwards at the same time as you measure future and past. Then you can slice up this combined space-time in different ways. A little bit like people slicing a shared piece of cake in different directions, giving themselves different amounts of chocolate and cream, or in this case, space and time, in the process.

The daydream ended abruptly when he reached the office. As he climbed the stairs to the third floor he shook his head to stop the mind from wandering. It was time for work, not play.

*

The young man could not possibly have known that his ideas would bring about a scientific revolution. The change would be dramatic. Yet, when it came, the revolution was not celebrated by newspaper headlines. There were no dancing in the streets. No talk about the dawn of a new era.

CHAPTER 1

The house on Mercer Street
1948 PRINCETON, NEW JERSEY

JACK HESITATED BY the metal gate. He couldn't help feeling a bit scared. He had walked past the white wood-framed house with the green window shutters many times, but he had never imagined he would one day go through the gate and knock on the door.

The house was not very different from the other houses on the street. It was a modest two-storey building with a narrow front, hidden behind a bulging hedge. It may not have been special, but the man who lived there certainly was. Albert Einstein. One of the smartest people in the world. And one of the most famous.

This was what made Jack nervous.

He had often seen the old man walk down the street. A living, breathing human being. But the boy's imagination was running amok. A mad collage of images flashed through his mind, fuelled by his avid reading of anything from the adventures of Jules Verne to his father's pulp magazines. The science of tomorrow in the form of fiction today. Exciting, mad science. Creatures from outer space. Alien invasions. Nuclear destruction. A giant calculating brain that decided to take over the world. Was this what was waiting for him? A shiver went down his spine.

He knew he was being silly. These were just stories. All in the imagination. There was nothing to be scared of. At the same time, he remembered his father explaining how Einstein's ideas had led to the bomb that ended the war. How could he forget? The teachers in school kept drilling them to duck for cover if the bomb came. He had always wondered how hiding beneath a desk would help against a weapon that could wipe out entire cities. But when he asked, he was just told that the way to stay safe was to prepare for the worst. This did not seem like much of an explanation.

He tried to stop his thoughts from spiralling out of control. Enough nonsense. He had a job to do.

*

Jack's father was a journalist at the Princeton Packet. His connections had helped the twelve-year old boy get the part-time job at the photographer's studio. So far the job had not been very interesting. The customers were few and far between so he spent most of his time tidying up. This was the first time he had been sent on an errand. It was a welcome change. He just wished the task hadn't been quite so daunting.

The photographer had sent him to deliver a set of pictures taken on the famous man's 70th birthday. Photos that would be sold to newspapers and magazines all over the world. The photographer wanted to make sure the professor was happy with them, but he had warned Jack that he had to take the reaction with more than a pinch of salt. "You see," he told the boy, "professor Einstein really doesn't like images of himself."

*

Jack had lived his whole life on Battle Road, a couple of blocks further along the route from the bustling main street. He had never met the famous professor, but he knew what he looked like. Didn't everyone? A long time ago his father had explained that the white-haired man who regularly walked past their house on his way to the Institute was Albert Einstein. The man that bent space and warped time. The man that explained that the tiniest speck of dust could harness enormous energies. The man that stuck his tongue out in pictures.

This had come as quite a surprise to Jack. Up until that point, he had thought the man was a caretaker or perhaps a gardener. How could this little man, with his baggy trousers, woolly jumper and a blue knitted hat pulled down over his unruly head of hair, possibly be a worldwide celebrity?

"You know," his father said, "you have to look at the inside. Don't judge the book by the cover. If you ever meet him, you'll understand what I'm talking about. Once you look beyond the messy halo of white hair and the sad brown eyes, there is a different quality. I think the word for it is genius."

Jack recalled this conversation when the photographer showed him the freshly developed pictures.

"It's not quite as it seems, Jack," the photographer said with a glint in his eye. "You see that coat? The way it's buttoned all up to his neck. Well, that's because he only had an undershirt on when I came to take the pictures. He didn't feel like putting on a shirt so he just grabbed that coat and did up all the buttons. It was kind of funny, but I think the pictures still look dignified."

"They do," Jack agreed thoughtfully, "but was he not expecting you? Did he not have time to prepare for the shoot?"

"Well, professor Einstein does things his own way. He has decided he's old enough that he can wear whatever he likes. He gave up on shirts and ties a long time ago. And he never wears socks, either."

"But," the boy hesitated before continuing, "that's a bit odd, isn't it?"

"Sure, but clever people are often quirky. You get used to it. Princeton is not a big town, but you'll find many deep thinkers here. Maybe more than anywhere else in the world."

He put the pictures in a brown envelope, sealed it and gave it to Jack. After checking that the boy remembered the address, he continued, "I've called the professor's assistant, Helen Dukas, to let her know you're coming. You can get going right away."

Jack moved towards the door, but the photographer carried on talking.

"Although," he said, "maybe you'd like to know how I first met the professor. It won't take long."

"It was more than ten years ago now," he started. "I was working as a freelance photographer selling pictures to various big city magazines. I was always on the lookout for a photo opportunity. One evening I was driving down Mercer Street as it was getting dark. Then suddenly, to my great surprise, I spotted a bodiless head at street level. It was just a head with two arms stretched above it and a strangely familiar mop of hair. Reacting on instinct, I grabbed the camera, jumped out of the car and snapped a picture. Only then I heard his croaking call for help. I rushed over, grabbed the man under the arms, and hauled out one of the world's greatest scientists. He had not paid attention to where he was going so had fallen into an open storm sewer."

Jack gave the man an astonished look.

"You may not believe it, but it's all true," the photographer assured him. "As I walked the professor the short distance to his house, he thanked me and begged me not to publish the picture. I promised I wouldn't and I never did. Actually, I couldn't have even if I'd wanted to. When I got home I discovered I'd forgotten to pull the film slide. It was a blank."

The man laughed and shook his head at the memory.

"Anyway," he finished, "it worked out well for me, because he's asked me to take many pictures of him since. Some have become rather famous. But I've held you up long enough. Now you need to go. Mustn't be late."

The walk from the photographer's studio on Nassau Street only took a few minutes. It was a familiar route. He usually took it on the way home at the end of the day. Heading west towards Trenton, he passed the seminary, where they trained the priests, and then reached the point where Mercer Street began to slope downhill.

*

He cleared his mind and pushed down the metal handle. The gate squeaked open. He walked across a small lawned area and climbed the five steps to reach the porch, passing the number 112 that was attached to the top step. He pushed the doorbell.

"Just a second!"

He could hear a muffled voice through the door. There was a rustling sound and then a woman with sharp features opened the door. After drying her hands on her apron, she gave Jack a firm handshake.

"You must be Jack," she said. "Pleased to meet

you. I'm professor Einstein's assistant. You can call me Helen."

She gave him a friendly smile.

"Don't worry. I don't bite. Come on in."

Jack stepped through the door into a different world. At first glance it was just a normal hallway, long and narrow with coats on hangers, boots and shoes on a rack, a couple of umbrellas in a stand. At the same time it was clear that this was not a normal household. Three large metal filing cabinets made the space feel cramped.

"I'm afraid the professor's papers have rather taken over the house," Helen said, noticing Jack's surprised look. "Anyway, let's take you upstairs. He's been waiting."

Jack followed her towards a spiral staircase at the far end of the hallway. Passing an open door on his left, he caught a glimpse of a large living room. The furniture was heavy, from a different time and place. A sharp contrast to what he was used to. The kind of room where ghosts might echo long lost conversations. He couldn't help shivering at the thought.

As they climbed the steep stairs, Helen turned to him and said, "I have to warn you. The professor is tired. He is recovering from an operation, so I have to make sure he's not working too hard."

There seemed to be nothing Jack could say, so he just nodded to show he had got the message.

"This birthday palaver," Helen continued, "has worn him out. He doesn't care much for birthdays. Other people make the fuss. He'd rather just keep on working."

They reached the top of the stairs. Helen knocked on a door to a room towards the back of the house. There was a gentle, "Come," from the other side. She opened the door and ushered the boy into Einstein's study.

CHAPTER 2

Machinery and mystery

THE ROOM SMELLED of dust, old papers and pipe tobacco.

As they entered, Einstein was seated in a diamond-patterned armchair, a blanket over his knees, a notepad on top of the blanket. He was working.

He put his notes down on a small side table and peered over the top of his glasses.

"You must be Jack."

The professor spoke softly, with a distinct German accent.

Awed by the moment, Jack lost the ability to speak. But he managed to nod in agreement.

"Have you got something for me?" the professor asked gently.

Jack handed over the brown envelope.

"Let us see," Einstein mumbled as he opened the envelope and riffled through the pictures. It did not take long.

"I hate my pictures," he sighed. "Look at my face. If it were not for this," he added, clapping his hands over his moustache in mock despair, "I would look like a woman."

Jack didn't know how to react, but Helen's

chuckle broke the spell.

"Oh professor," she laughed, "that's nonsense talk. They look fine. And you certainly don't look like a woman."

The professor turned to Jack. Suddenly he seemed much younger, almost like a cheeky schoolboy.

"What do you think, young man?" he asked. "Do you agree with Die Dukas?"

Jack had no idea what to say. If he agreed, it might seem rude, but on the other hand, disagreeing did not feel quite right either.

"Uh, well..." he spluttered.

"Or perhaps you would rather talk about something else?"

The professor smiled.

Jack had taken the opportunity to look around the room while Einstein was flicking through the photos. Two of the walls were covered with floor-to-ceiling bookcases. Every single shelf was stacked with books and papers. In the middle of the room stood a large table, covered with pencils, papers and a number of pipes. The far end of the room was dominated by a large window, looking out over the back garden. On the windowsill lay a gnarled tree branch. There was an old globe in the corner.

Desperately trying to find something to say, Jack noticed portraits of two somber looking, possibly Victorian, gentlemen on the wall. One of them had a bushy beard, the other was clean-shaven and looked much younger.

"Uhm, who are they?" Jack stuttered as he gestured towards the portraits.

"Ah," the professor responded, "Maxwell and Faraday. Great thinkers. The founding fathers of electromagnetism. What do you know about electricity and magnetism, Jack?"

"Well..." The boy thought for a moment. "We've learned about Ben Franklin and the lightning."

"Good old Benjamin Franklin," Einstein commented, "a great man. He is an important part of the story. What else?"

"I guess there was Tom Edison," the boy added, "and the light bulb."

"Ah yes, Thomas Edison. The inventor," Einstein replied. "Those two were American. The men on the wall were British. They explained how electricity and magnetism actually work. In fact, they showed that the two are close relatives. Electricity can become magnetism and the other way around. Like in an electric motor."

"Michael Faraday is probably my favourite," the professor continued. "He was not the best mathematician, but he had great intuition. He loved the mysterious part of nature. Back in his day, scientists had the freedom to think about whatever they wanted. There were not these dull specialists we have today, staring at the world through their horn-rimmed glasses. Destroying the poetry. Faraday was different."

"Of course," he added, "it was the other man, James Clerk Maxwell, that came up with the theory behind it all. He showed that the speed of light is fixed. No matter how you measure it, however you move, you always get the same answer. The mathematical equations he wrote down were beautiful..."

The professor noticed the frown on Jack's face.

"But maybe you do not think mathematics can be beautiful?" he asked.

The question made Jack uncomfortable. He had never got along with mathematics. The truth was that he found it boring. Dull and complicated. He had certainly never thought of it as beautiful.

"Anyway," the professor saved the boy from having to figure out an answer. "I have battled their creation all my life. Sometimes it feels like I am winning. Other times... well, let us not talk about that."

"Now, professor, I'm not sure Jack has time to listen to your old stories," Helen interrupted.

She gave the boy a look, as if offering him the opportunity to leave, but he had to decide for himself. She knew that, once he got started, the professor wouldn't stop. He loved talking about his ideas.

Jack knew he ought to get back to work. He had delivered the pictures. He'd had the reaction. Not great, but that had been expected. Now it was time to leave. But the professor's story had stirred something inside the boy. Something he hadn't known was there. He had never been interested in science. At least not the kind of science they were taught in school. An endless list of facts and figures to remember. Not even the teachers seemed to care. This scruffy-looking, old man intrigued him. His science seemed different. Like an adventure story with an explorer leading an expedition into an unknown territory, reporting back his discoveries, explaining not just what but why, as well.

Jack wanted to hear more.

Helen gave the boy another look. When he didn't make a move, she made the decision for him.

"Well," she said, "if you're going to stay, I think you should have a seat."

She motioned him over to an armchair.

"I'll go back downstairs to prepare your lunch, professor. Don't keep Jack too long."

She almost added, "and don't confuse him too much," but stopped herself when she noticed the beaming grin on the professor's face. He needed something to cheer him up and if the boy ended up confused, then so be it.

"Shall we continue?" the professor asked as the door closed behind his assistant.

Jack nodded.

"Let us see," the old man cleared his throat. "It is a long story. I remember, or at least I think I remember, it was many years ago. I was a little boy, maybe 4 or 5, and I had been ill for several days. My father brought me a compass to cheer me up. I was fascinated by it. No matter how I twisted or turned it, the needle always pointed in the same direction. I tried to trick it, but it was too clever. I had no idea why it was behaving in such a stubborn way. How could it be? Later, my father explained how it worked. He told me that the earth is like a big magnet that makes smaller magnets, like the needle in the compass, want to point north at all times."

After thinking for a moment, he added, "It was still a bit mysterious, I guess, because I could not understand how the magnets could affect each other. I only had experience with things moving because they were pushed. The needle of the compass was isolated in its little chamber. Nothing was pushing it around. This invisible communication puzzled me for a long time, but I learned something important. I learned that something deeply hidden determines how things work."

"When I got a little bit bigger," he continued thoughtfully, "my father used to take me along to the factory he was running together with his

brother. It was an exciting place, full of machinery and mystery. They made all sorts of electrical devices, dynamos and generators. I really enjoyed trying to figure out how the different contraptions worked."

"Unfortunately," he added, "my father was not a very good businessman. The factory had to close."

The professor went quiet, lost in the memories.

A couple of minutes passed.

Jack started feeling uncomfortable just sitting there, watching the old man think.

"But what about the men in the pictures?" he asked hesitantly to break the silence.

"What?" the professor reacted with a start. "Oh, them. I am sorry, I forgot about them. They came later. Or, actually, before."

"I did not know much about Maxwell or Faraday until I became a student," he continued. "The professors at university were keen on explaining the ins and outs of electromagnetism, but I did not always pay attention to what they were saying. I might have been too interested in my own ideas. I wanted to do my own experiments. I could not help feeling there was something missing... But maybe I am going too fast."

The old man paused for a moment before continuing.

"Maxwell's mathematics explained that there should be electromagnetic waves. Waves that move with the speed of light. This was important. It led to many great inventions. Like the radio." "You know," he said quietly, as if sharing a deep secret, "I worried a lot about light when I was young. I used to ask what it would be like to ride along a beam of light. What would happen? What would the world look like? This made me think about the mystery of time. Is time like a river that flows downstream, from the past to the present, or do all times exist together?

What do you think, Jack? Do you think we can break the flow of time? Or would that make us remember things that have yet to happen?"

Jack was stunned by the dramatic turn of the conversation. One moment they had been talking about two men from the past. The next moment he'd been asked if he could remember the future. He was confused. What could he possibly say? It felt like the professor was trying to trick him with his complicated logic, but he couldn't say that. Could he?

An old grandfather clock chimed downstairs. The awkward moment was gone. It was noon. The old-fashioned tyranny of time reminded the professor of lunch.

They went downstairs, where Helen was waiting.

As Jack was about to leave, the professor showed him a wind-up toy he'd been given for his birthday. It was a funny-looking bird with suction cups on a rotating wheel instead of feet. The old man put the toy against the hallway mirror, pulled the string, and chuckled as the bird climbed all the way to the top.

Jack was astonished. The famous Albert Einstein was playing with a kids' toy.

The professor watched Jack's face the whole time. Intently.

"Did you like that?" he asked.

"Yes," Jack said, without the slightest hesitation. He did like it.

Helen walked him to the door. After saying goodbye, she added, "You're his friend now. If you want to come back and visit again, you're welcome. I can tell that he enjoyed talking to you."

CHAPTER 3

Things fall into place

THE WORLD'S MOST celebrated thinker lay on a small cot as tiny metal electrodes were attached to his scalp, up his nostrils and against his eardrums. He was acting as guinea pig for an experiment aimed at finding out what mechanism in the brain allows a genius to think through problems that are too complicated for a normal person.

Having finished wiring the professor, the lab assistant switched on the electronic equipment. He stood back as the machine charted the brain waves, measuring the faint electric currents that pulse through the brain, recording them for detailed analysis.

*

"So what did they say?" Jack asked the professor later that afternoon.

Over the last couple of months, the boy had become a regular visitor to the house on Mercer Street. He enjoyed the professor's stories almost as much as the professor enjoyed telling them. It may have been an unusual kind of friendship, but it worked for both of them.

Jack had come over to help trim the hedge in front of the house, and now they were sitting at the long table in the dining room.

It was a chilly day. A fire was crackling in the fireplace. The room was gloomy as the greenery outside the windows filtered out the sunlight. The darkness gave a sense of the true age of the house, a sense that was strengthened by the wide, sagging floorboards and the ceiling that dropped gradually towards the back of the room.

"Ach," Einstein responded with mock disgust, "they have no idea what they are doing. They are just playing games."

"Yes, but tell us, professor," Helen implored. "How different are you?"

For some reason she seemed amused by the whole thing.

"I will tell you what they say," the professor gave in. "They say their electronic machinery shows them my brain is different. They think my mind is not the same as that of an ordinary person. They claim my thinking tunes in on groups of brain cells in quick succession, scanning the entire brain for the correct answers. The way a radar antenna searches the sky for aeroplanes."

"I think," he added, "it is nonsense. And I expect an ordinary person would agree."

"But surely," Jack hesitated for a moment, trying to find the right words, "surely you're not an ordinary person? I mean, you do seem to think differently."

"Ach" the professor exclaimed again, "it may

seem to you that I think differently, but maybe I just give the ideas more time to grow."

"Let me give you an example," he continued. "What can you tell me about gravity, Jack?"

The boy hesitated. He was getting used to the way the old man liked to turn conversations around. Putting people on the spot. Making them think.

"Well," he replied thoughtfully, "I know about Newton and the apple falling on his head. And I know that gravity is good because if we didn't have it we'd float away into space. And I don't think there's gravity in space."

The professor gave him an amused look, as if he had been expecting more.

"Ah yes," the boy added, "I believe there should be something about warped space, as well. Something you came up with. But I have no idea what that means."

Einstein nodded approvingly.

"Then let me tell you a story," he said. "When I was a young man I worked in the patent office. It was a good job. I had fun thinking through other people's inventions, trying to figure out if they would really work. It was interesting, but once I got used to it the work was easy. I had a lot of time to think about my own ideas."

"You know," he chuckled, "I used to have my calculations hidden in the desk drawer, working

on them whenever I thought nobody would notice."

"That seems a bit naughty," Jack commented. "Like not paying attention in class."

The thought of the school's strict classroom discipline sent a shiver down his spine. The smartly dressed teachers, with neatly pressed shirts, jackets and ties, made sure everyone knew who was in charge. In no uncertain terms. Jack had learned this the hard way. He was often caught daydreaming.

"Yes, it may have been a bit naughty," the professor agreed, bringing Jack's attention back to the conversation. "But I did my job well so there were no complaints. Then one day, maybe I was bored, who knows, I was staring out the office window. That was when I had the idea. It may have been the best thought in my whole life. I imagined seeing a man falling off the roof of the next-door building. It struck me that this man would not actually feel the effects of gravity as he was falling."

"But," Jack interrupted with a look of concern on his face, "what about when he landed? That would hurt, wouldn't it? Wouldn't this mean he did feel the effect of gravity?"

"Obviously," the professor conceded, "but he would not feel gravity as he was falling. Even though he was accelerating towards the ground. That is the point."

"This simple thought experiment made me see that you can not tell the difference between gravity and acceleration," he continued. "You would not be able to tell if the man was falling past the window because of gravity or if it was me and the desk that were accelerating upwards. This was important. It gave me a clue to how gravity actually works."

Jack frowned.

"I don't get it," he said, shaking his head slowly. "Doesn't everyone understand how gravity works?

It makes things fall. What else is there?

"That is a good question," the professor agreed, "but there is a difference between understanding what something does and why that should be so. The first question can be answered by an experiment, the second requires a theory."

"A long time ago," he reflected, "when people thought the world was flat, they had no reason to think about gravity. There was up and down and that was it. It seemed natural that everything should move downwards, and no one thought to wonder why. Then Newton came up with his theory. Apple or no apple. He said gravity was a force that attracts everything that has weight, weakening as the distance increases. This was a great success. Newton explained why the planets move in orbits around the sun, why the moon goes around the earth, why there are tides in the oceans and so on."

The professor paused for a moment to gather his thoughts, checking that Jack was still paying attention.

"Newton's theory explained a lot. But then I managed to break it..."

It was a strange confession and Jack did not know how to react. But he didn't have to. The professor picked up the thread.

"You see, Newton's gravity was like magic. Lift a finger, and the change in gravity would at once, in absolutely no time, be felt throughout the universe. Then there was my theory of relativity, where nothing could move faster than light. Not even changes in gravity. So Newton had to be wrong."

"That was when I thought of the falling man," the professor smiled, "and the pieces started coming together. I asked myself what would happen if I were in an elevator accelerating upwards at a

tremendous rate. The acceleration would push me towards the floor, acting just like gravity. As I imagined approaching the speed of light, I realised that a beam of light entering through a hole in the wall would appear to bend down in an arc towards the floor because of the acceleration. The light moved at its fixed speed across the elevator but the elevator itself was moving upwards. Of course, to me, riding in the elevator, the effect would be the same as if gravity were bending the beam."

"There you go!" He laughed, waving his arms in the air like a magician that had just pulled a rabbit out of a hat. "Gravity bends light!"

"In the end the idea is quite simple," he finished. "Light is lazy. It wants to move in a straight line. Yet gravity makes it bend. One way to explain this is to make space curve. This way you can allow light to be lazy and follow the shortest path. But it took an awful lot of hard work to put the mathematics together. Nearly ten years."

They carried on talking late into the night. Eventually, Helen interrupted them.

"Young man, you're keeping the professor awake," she scolded Jack.

"I do not think so," the professor replied. "It is my fault. I have been keeping Jack awake."

*

At breakfast the following morning, Jack's mother interrogated him. She had noticed that he had got back late. Again. She knew where he had been, but she was not sure how she felt about it. She obviously knew her son was special, but Albert Einstein was in a different league. Was it really good for her Jack to spend quite so much time in the company of that weird old scientist? Not to mention the harsh looking housekeeper. But she had noted how, all of a sudden, Jack seemed to be doing better at school. Something had changed.

Even though she could not make her mind up what she thought, she still had to ask where he'd been and why he'd been back so late. It was a school night, after all.

"I'm sorry I was late, mom. Completely forgot the time. Professor Einstein was explaining his gravity theory," Jack told her. "It was interesting, but really confusing."

"I got nowhere trying to grasp what he wanted me to," he admitted ruefully. "At the end of the evening he'd filled a whole pad with weird symbols. Like some outer space maths. I had no idea what was going on."

"But now that he's explained his theory to me, there's one thing I know," he looked up from his breakfast cereal with a grin. "He sure understands it."

CHAPTER 4

Light askew

JACK WAS SWEATING. It was a sweltering hot summer day and there was not even the slightest breeze to provide relief. The humidity didn't help, either. He was lying on his knees with his head buried in the flowerbed. Pulling up weeds. Little pieces of greenery determined to grow wherever they were least supposed to. He was tired and his back was aching.

An annoying bead of sweat slid all the way down the boy's forehead and launched itself from the tip of his nose. Caught by the grip of gravity the droplet fell towards the ground. Jack sat up and wiped his brow, letting out a heavy sigh. He was in desperate need of a cold drink.

"Great job, Jack!" Helen called from the top of the veranda. She was coming back from the kitchen, carrying a tray with glasses and a jug of freshly made lemonade. Ice cubes clinked against the glass jug. The sound made Jack's mouth water. *She must be a mind reader*, he thought, already feeling a lot better.

He had not quite finished tidying up the decorative flowerbeds at the back of the garden, next to a tiny vegetable patch. For some reason the professor was extremely proud of this vegetable patch, but Jack could not see much evidence of anything actually growing there.

The professor was sitting on a deck chair in the shade provided by one of the many trees that dotted the lawn. Even though he was in the shade, he was wearing a funny-looking floppy hat. The wire-haired terrier, Chico, was sleeping by the side of the professor's chair. On the other side there was a small folding table with a pile of letters.

When he sat down to start his reading, the professor complained that the daily mail took up so much of his time. He used his hands to show how thick the bundle of letters tended to be. Then he scratched the dog behind the left ear, and said, "At least you feel sorry for me, my friend. I think that is why you always try to bite the mailman."

The dog gave the professor an earnest look, then curled up to go to sleep next to him.

A man from the local astronomy society was busy mounting a small telescope in the bright sunlight. It was to be a gift to a school in Israel, but the professor wanted to see it working before it was shipped. In between observing the construction and asking the occasional question, he flicked through his letters. Now he was just sitting with his eyes closed. He might be asleep, or perhaps deep in thought. It was hard to tell.

The man groaned as he tightened a nut on the telescope stand. He was soaked in sweat. He took his hat off, used it to fan his face and then wiped his damp neck with a handkerchief.

"I do believe you're about to save my life, madam," he sighed with relief. "That lemonade looks delicious."

"Anyway, this contraption is ready for inspection," he added a moment later. "Although you won't see much of the stars as long as the sun is out."

"It does not matter," said the professor, apparently awake after all. "I do not think anyone here knows much about the stars. I certainly do not. When it comes to astronomy, I built nothing but a spacious castle in the sky. The constellations have always been a mystery to me."

"But it is a common misunderstanding," he continued as he got out of his chair, walked over to the telescope and peeked through the eyepiece. Apparently satisfied, he nodded and returned to the chair. He picked up a letter from the pile and started reading.

"Just listen to this," he said. "Dear Professor Einstein. My father and I are going to build a rocket and go to Mars and Venus. We hope you will come too! We want you to go because we need a good scientist and someone who can guide a rocket good."

The professor paused for a moment. Then he let out the most extraordinary kind of laugh. It was rather like the barking of a seal. A happy laugh.

"Imagine that!" he exclaimed with a broad smile. "Mars and Venus. I can not wait."

"Although..." he continued, "it may not be such a good idea. My sense of direction is not great. I sometimes even get lost on the way back from the office. And when it comes to the stars, well, it makes me particularly happy when they are found in the wrong place."

He took a sip of lemonade and gave the man a thoughtful look.

"Maybe you would like to know why?" he asked.

Jack knew what was coming. The professor loved to explain his theories and he had heard this particular story many times. He still didn't quite understand it, but he enjoyed it nevertheless. The professor had infinite patience and a contagious enthusiasm. He was a good storyteller.

"Well," the man replied with a worried look on his face, "if you're going to try to explain that relativity thing, then I don't know if I do. As far as I understand it, your theory is not very well understood."

"In fact," he added, "my son has read your book, and he tried to explain it to me. The only thing I picked up was that everything is relative."

"Oh... and space sags like a mattress under a fat man," he finished and gave his round belly a couple of satisfied pats.

"Mattress... what mattress?" asked the professor with a frown.

"See, that's what I'm talking about," said the man. "You're confused, as well, and you haven't even started talking about it yet."

He sat back in his chair and took another sip of lemonade, but if he thought this was the end of the matter then he was mistaken.

The professor hesitated for a moment. Then he sighed, "The meaning of relativity has been widely misunderstood. Everyone plays with the word, like a child with a doll. But relativity, as I see it, merely describes certain physical and mechanical facts, which are relative to one another. It does not mean that everything in life is relative and that we need to turn the whole world topsy-turvy."

"But I know better than anyone that the theory is difficult to explain," he considered, getting into his stride. "Everybody thinks they know what space means, and they also know what a warp is. Yet, hardly anyone understands the meaning of a warped space. The words are so familiar you get the impression that you have been let in on some deep and meaningful secret. Be that as it may. The difficulty people have in understanding these ideas is nothing compared to the difficulties I had putting them together."

"I guess it does not help," he finished, "that I had to topple my own house of cards to build a new one. If my first theory was strange, with clocks running slow and moving objects seeming short, then the second one was truly peculiar. Gravity makes time hesitate and light rays bend. A straight line is no longer the shortest path between two points. As far from our everyday experience as you can get. Yet, a lot of things work out if you let gravity curve space and warp time. It may be strange, but it fits."

"Hrm," the man cleared his throat to interrupt the stream of information. "It sounds grand, but why does it have to be so complicated? And what do you mean by curved space, anyway?"

Jack had listened intently. At this point he decided to join the conversation. He knew exactly what was coming.

"You know," he said, "I struggle to understand this as well, but I'm beginning to see how it works. Whenever matter is present, where there's a mountain, an earth, a sun or where a comet moves, space sticks out a little bit more than where there's emptiness. In turn, this change in the shape of space affects how things move through it."

He picked up a twig from the ground and bent it gently before continuing, "Imagine a blind beetle that crawls along the surface of a curved branch. It doesn't know the track it has covered is bent."

The professor nodded his agreement.

"You are right, Jack," he said. "I was lucky enough to notice what the beetle did not notice."

"But," the man interrupted, "I still don't get it. What does it mean? What does space stick out into? It doesn't make any sense."

"Ach," the professor grunted, "you need to understand the different dimensions of the problem."

"Let us imagine being as small as Jack's beetle," he continued, "limited in sense and understanding to a single dimension. The world it lives in would have to be a line. Now, if this line were curved, returning to itself like a circle, then the beetle would never reach the end. So it would think the world must be infinite. Without an inkling of the existence of other space dimensions, and unable to understand the nature of the surface on which its circular world was drawn, it would be unable to figure out that its world only appears to be endless."

"I'm not sure what a beetle on a stick has got to do with anything," grumbled the man. "I thought we were talking about the mysteries of space and time."

"That's just it," said Jack. "A two-dimensional being would have a similar experience. Imagine being flat like a piece of paper, with the imagination and intelligence to understand only two space dimensions and living on a two-dimensional surface like a ball. Then you could never arrive at the end of the ball, and would have to be convinced that the world is endless."

"Right, Jack." The professor encouraged the boy to continue his train of thought.

"Now take a real human being, capable of seeing, understanding and imagining all three dimensions. Up and down. Left and right. Backwards and forwards. The way the world appears to us. A fourth space dimension is unknown to us, may seem impossible. What if such a fourth dimension existed anyway, if our space were also some such thing as a circle or a ball? Something curved, perhaps returning into itself. Then maybe this warping of space could explain why the planets move the way they do. They're trying to move in a straight line, but the space itself is curved. That's the professor's theory." Jack finished his explanation.

"My, oh my," Helen chuckled. "I do believe the professor's talking has rubbed off on you."

"Maybe," said Jack with a shrug, "I'm beginning to have some idea of the words. But the professor works things out with mathematics, balancing the weight of the cosmos on the tip of his pencil. I'd never be able to do that. Not in a million years."

"The mathematics is important," agreed the professor. He scribbled something on the back of one of the envelopes from his pile of letters. Holding up the result, a neat little formula, he said, "Be sure you take a good look at them. They are the most valuable discovery of my life."

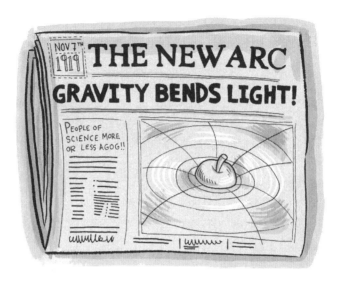

"You need the equations," he continued, "because they allow you to make predictions. To test if the idea works. If a single one of the conclusions drawn from my theory proves to be wrong, then the theory falls entirely. I do not think you can modify it without destroying the whole thing."

"So far, I have been lucky", he added. "All experiments agree with me. But it could have been different..."

He leaned back in the chair, closed his eyes and thought for a moment.

"In the end, it had to be the British," he reflected. "If Newton's old law of gravity had to fall, it made sense the British should be the ones to prove it."

*

The calculation of the light bending was wrong in the first version of the professor's theory, and several quirks of history saved him from the embarrassment of this mistake.

Astronomers listened when he explained that one ought to be able to measure how the light from distant stars was deflected by the sun's gravity during a solar eclipse.

There were a number of attempts, but it was not until after the war, when the professor had corrected his calculations, that there was a successful measurement.

CHAPTER 5

The power of the sun

GROWING UP IS the easiest thing in the world. You don't even have to think about it. It happens anyway. But this does not mean it is always easy. Being a teenager can be particularly difficult. When you are little, you don't have to take responsibility for anything. This changes as your grow up. At some point, you don't seem to get away with things any more. People expect you to be in the right place at the right time. All by yourself. And you are supposed to do the right thing, as well. All the time. This is when you realise you are no longer a child. You have become a teenager. Of course, this does not mean people won't treat you like a child. They will.

*

Jack did not find being a teenager particularly easy. It was confusing. He had no idea what was expected of him. His mother kept calling him "her little boy" in front of perfect strangers. This was intensely embarrassing. He was, after all, nearly as tall as her. At the same time, she kept asking him to help out around the house. Especially when he was busy doing something he really enjoyed. This was frustrating.

He was trying to figure out who he was and

what he was supposed to be doing, but not making much progress on either question.

School added to the frustration. It was getting serious. All work and no play. He had to pay attention all the time.

The fact that his friends were also trying to figure out how to be teenagers made life complicated. People that had been friends since they were little fell out, often for no apparent reason. New friendships were forged, but these were often fragile and could be over before they really got going.

Jack tried to keep his head down and stay out of trouble. On the whole, he managed to remain invisible. One of the crowd. It happened that people tried to make fun of his friendship with "that old coot" Einstein, but Jack found it easy to ignore those comments.

As Jack became more aware of what was going on around him, he realised there was a bigger problem. The whole world seemed to have an identity crisis. Ever since the Soviets exploded their first atomic bomb, the political climate had been frosty. Everyone was afraid there would be another war. And this time both sides would have terrifying weapons. Weapons that could wipe out entire countries, in a single moment. Politicians saw enemy spies everywhere. People stopped trusting their neighbours. Everyone was a suspect. Everyone was scared.

The professor was in a dark mood.

"You know, Jack, it really is quite simple," he said as he straightened up his tie for the third time. "Mass is energy and energy is mass. My little equation says they are two sides of the same coin. This makes a lot of sense. You know sunlight carries energy because it makes you hot. If this energy has weight, then light should obey the rules of gravity. It is perfectly logical. The light from the stars should bend. And we know it does."

"But energy comes in many forms," he added, "and they can change into one another. The energy of movement, like in the flow of a river, can turn into electricity through a generator. The electric energy can generate heat via a radiator. Then there is nuclear energy..."

He paused.

"Nuclear energy," he repeated sadly. "The power of the atom. The ability to release the extraordinary amount of energy that is stored in a tiny bit of dust. My equation made this possible. And I am not happy about it."

*

Jack was astonished by the professor's transformation. When he arrived, the old man had greeted him on the porch dressed in baggy trousers and a bedraggled sweater. He'd been wearing sandals, as usual. But he had shed the baggy clothes and put on a dark grey suit. He was wearing shiny black shoes. He may not have been wearing socks, but he still looked very smart.

The professor took a look in the mirror. Adjusted the tie a final time.

"That will have to do," he decided.

He gave the boy a wry smile.

"Are you ready to go, Jack? No need to be nervous, although... I hear even the ambassador of Mars will be there."

The doorbell rang.

"That must be Oppenheimer," said Helen, emerging from her office.

The dog sauntered over to the door. Lazily, as if he almost couldn't be bothered, he made a sound somewhere in between a growl and a bark.

"Poor Chico," Helen bent down to scratch him behind the ear, "I do believe you're a bit bored."

As she turned towards the door, she added, "You know, professor, you and the dog are just the same. Energetic and intelligent, but with a low threshold for boredom."

She opened the door to greet an intense, nervous, gauntly thin man wearing a brown coat and a porkpie hat.

"Evening, Helen," the man said as he took off his hat. "Evening, professor."

"And you must be Jack." The man turned his icy

blue eyes towards Jack. The boy recognised the thin angular face and the close-cropped hair from pictures in the newspapers.

Robert Oppenheimer. The scientist that led the Manhattan project. The man that built the bomb.

*

Jack had never really considered the possibility that the professor's thinking could do any damage. How harmful could warped space and wonky time possibly be? But when the professor told him that Oppenheimer would be joining them for the evening's official dinner, the conversation had inevitably taken a darker turn.

The professor explained that he had mixed feelings about his neat little equation that linked energy and matter. On the one hand, he was proud of it. "Imagine the audacity of this step," he said. "Every clod of earth, every feather, every speck of dust became a reservoir of trapped energy."

"Of course," he continued with regret, "there was no way of verifying this at the time. This did not happen until more than thirty years later, when German scientists split the atom. Just before the war. I never would have thought of it, but some of my colleagues warned me that one might be able to create a nuclear chain reaction. This was when I wrote to the President, to warn him about the possibility of an extremely powerful weapon."

"I may not have had anything to do with the bomb project," he added, "but not a day goes by that I do not regret writing down that equation."

*

Now Jack was being introduced to the man that unleashed the power of the sun. He felt a chill as they shook hands.

The professor had warned him that Oppenheimer was a troubled man.

"He carries more top secret information in his head than any other man alive," he explained. "There should not be the slightest doubt about his loyalty, yet they are treating him like a villain. It is a witch hunt. They have revoked his security clearance. It is a great shame that a man that has done so much for his country should have this done to him. It does not make any sense. But everything, even lunacy, is mass-produced in this country."

*

After saying goodbye to Helen, Oppenheimer led Jack and the professor through the metal gate to a black car parked by the curb. The driver that was waiting for them was entirely dressed in black. His eyes were hidden behind dark glasses. The sinister appearance left little doubt that this man was more than just a driver.

As the car pulled away, Oppenheimer looked out the window. After a couple of minutes, he

51

started talking. There was something distant about his soft voice.

"It's kind of funny," he said. "Every time I visit your house, the dog reminds me of the past. I used to have a horse called Chico, you know. I loved riding. Used to go exploring the rugged mountains out west. In the evenings I made camp, read by the light of an oil lamp and slept under the stars. I even took my students on some of those hikes. We talked about everything and nothing. I miss those days. I really enjoyed the teaching."

"And I think," he added, almost as an afterthought, "I think I was a good teacher. The students used to joke that I knew the answer even before they figured out the question."

He turned to look directly at Jack and the professor. The blue intensity of his eyes reflected his restless energy.

"I do not consider myself a teacher," the professor admitted. "I am still a student. But I share what I know with others. I am also not very organised. Definitely not a practical man. All my life I have worked with my head. I have never built anything with my hands."

There was a shadow of a smile on Oppenheimer's gaunt face.

"I know what you mean, professor," he said. "I can tinker with a car, but not very well. Thinking suits me better."

The two men started talking about their ideas. The professor described the progress on his latest attempt to explain matter in terms of space. Oppenheimer's penetrating eyes followed intently. The tips of his long fingers joined in thought.

As they arrived at their destination, the professor interrupted the gentle flow of the conversation.

"Robert," he said, somewhat abruptly, "you know what you have to do. You have to go to Washington. You need to tell the officials they are fools, and then go home."

"There are days," Oppenheimer replied, "when I would like to do just that. But I don't think it would be very practical. Things are complicated enough as they are."

*

The next couple of hours passed in a blur, with a host of smartly dressed people mingling and talking, clinking champagne glasses. Famous faces blended with less familiar ones. Jack was still young enough to stay invisible. But he noticed how everyone crowded the professor, and he was not at all surprised when the old man leaned on him for support.

"I'm glad it is over," the professor sighed, when finally it was.

The professor took a deep breath as he sank back into the car seat. It had been an exhausting evening. He closed his eyes, but it was clear that he could not relax. In a quiet voice, he explained that he kept mulling over the equality of energy and matter. The tragic aspects of his theory left him devastated.

"It weighs heavily on me," he said. "I feel responsible for the death of all those thousands of human beings. All gone in an instant."

Oppenheimer looked at him. Sad and serious at the same time.

"You're not to blame for this madness, professor," he said. "To say that your equations made possible the construction of these weapons is like saying that the invention of the alphabet caused the Bible to be written."

"If anyone is to blame, it is me," he sighed. "But I'm not sure about that either. Us scientists are a slow, plodding lot who are painstaking in our experimentation. We don't work well with timetables. It was the generals that set the deadlines, demanded results and refused to accept excuses. It was the generals that made it happen. We were just the tools."

"The project itself was an adventure," he admitted. "I don't think we were some kind of nuclear playboys, exploding these bombs to satisfy some personal whim. But the secrecy of it all was tremendously exciting. And if I'm honest, I'm not sure we actually expected the bomb to work."

"That first test changed everything," he continued. "I remember it vividly. As the last seconds ticked away, I could scarcely breathe. Then the announcer shouted 'Now!' and there came this tremendous burst of light, followed by the deep-growling roar of the explosion. A few people laughed, some cried, but most people were silent."

"I am become death, the destroyer of worlds," he whispered, leaving the car in silence.

"There was no need to ever use it again," the professor said after a while.

"No, there wasn't," Oppenheimer agreed.

*

It was late when they pulled up outside the house on Mercer Street, but the lights were still on. Helen threw the door open as Jack was helping the tired professor up the stairs. She was visibly ruffled.

"What is going on?" asked the professor. "I did not expect anyone to be up at this hour."

"There was a telegram," stuttered Helen. "The president of Israel is dead."

This was disturbing news, but the professor could not see why Helen found it so upsetting.

"What..." he started, but Helen interrupted him before he got any further.

"They want to offer you the presidency," she gasped.

The stunned professor shook his head gently, in

obvious disbelief.

"This is awkward. Very awkward," he muttered.

He started pacing up and down in a state of agitation, which was very unusual for him.

"It is madness," he insisted, "complete madness. I am not a people person. I have neither the ability nor the experience to deal with human beings. How can they possibly expect me to do this? What will they say when I say no?"

"That's not what worries me," said Helen. "I'm more worried about what happens if you accept."

CHAPTER 6

A piece of junk
1954 PRINCETON, NEW JERSEY

TIME PASSES. DAYS become months, months turn into years. The past turns into the future. Tomorrow becomes today. Always that way around. Today does not become yesterday. At least not until tomorrow...

Time flows in one direction. Without negotiation. It's a one-way street. Different people may experience time differently. They may not agree on what time it is. But everyone agrees that time flows towards the future. That's just the way it is.

*

Jack had never really thought much about time until he met the professor. He had read about time travellers wreaking havoc with past and future, but never taken this too seriously. They were just stories. He had not asked himself if one might actually be able to travel in time. Then the professor confused him by explaining that they were travelling in time all the time. Into the future. One second every second. When Jack protested, the professor just mumbled something about not violating causality. Nothing happens without something making it happen. That gave Jack a headache.

*

It was the last week before Jack was leaving for college. He had been busy saying goodbye to friends and preparing the journey. He had bought a beaten up Volkswagen Beetle and was planning to drive it all the way across the continent. His mother was not happy with this idea. She kept referring to it as a hare-brained scheme and tried her best to talk him out of it. But Jack was excited. He thought it sounded perfect. What better way to start his new life than with a road trip?

As he tried to finish the packing he heard the phone ring. A moment later, his mother called up the stairs to tell him to pick it up. It was Helen. The sound of her voice gave Jack a twinge of guilt. He had not visited the professor for a couple of weeks and soon he would be gone for a long time.

"The professor is sulking," Helen told him. "Doctor Dean has forbidden him from sailing. Says he's not well enough. He's very unhappy, Jack. It's really quite sad. Seldom did I see him in so light a mood as in his strangely primitive little boat. I was thinking... I know you're busy, but... I was thinking you might be able to take him out on the lake before you leave. It wouldn't have to be long. Just something to break the routine."

Carnegie Lake, thought Jack, *should be nice on a day like this*. The answer was obvious.

"Of course, Helen, I'd love to do that. I'll come by after lunch."

"That's fantastic, Jack. I'll tell the professor you're coming. That will cheer him up. Shall I arrange for a car?"

"No need," said Jack proudly, "we can take mine."

There was an "Oh" followed by silence at the other end of the line. Then Helen recovered from the surprise and finished, "See you soon then, Jack. Goodbye."

Jack put the phone down and carried on packing.

*

He arrived at Mercer Street a couple of hours later, climbed the familiar steps and rang the doorbell. Helen opened the door, but before she had the chance to say anything Chico barged her aside and jumped up to welcome Jack.

"At least you've missed me, eh?" laughed Jack.

"I think so," Helen agreed. "The poor thing has been sulking alongside the professor."

"I'm afraid you can't come along where we're going," said Jack gently, giving the dog a good scratch. "Where's the professor hiding, anyway?"

"He's in the music room, tinkling on the piano. He's given up the violin. Says he can't stand the noise he makes on it. He's not happy, Jack. Something needs to change. A trip on the lake should do him a world of good."

"Ah, Jack!" the professor exclaimed as the young man entered the room. He looked up from the piano. His hair was as wild as ever, but the face had a tired look. The brown eyes had lost their spark. "There you are. Wonderful to see you."

"You look well, professor," Jack replied, even though it wasn't true.

"I doubt it, my young friend. These days I stay in bed and receive guests like an old lady of the eighteenth century. This was fashionable in Paris at the time. But I am not a woman, and this is not the eighteenth century!"

The soft voice could not hide his frustration.

*

They arrived at Carnegie Lake half an hour later. The professor had chuckled at Jack's car, suggesting it was a chariot suited for kings and emperors rather than old professors, but he seemed to enjoy the ride. He kept the window rolled down to let the fresh late-summer air stream through the car. He was wearing a wide-brimmed hat, a baggy shirt and a pair of well-worn linen trousers. Jack noticed that they were held up by an old bit of rope.

As they walked around the corner of the boathouse, they saw an elderly couple talking to one of the boatmen. Jack heard the professor take a deep breath as he spotted them. They were clearly familiar to him. There was a moment of hesitation before the professor smiled and said, "Good afternoon."

"Ah, my dear friend professor Einstein," the man replied, with a heavy Russian accent. "Did you have a chance to read my book?"

"Ach," exclaimed the professor, as if something had slipped his mind, "I am afraid I have not managed to read it yet."

"I would like an occasion to meet you and discuss," continued the man.

"But what do you really know of astronomy?"

"I know to put questions."

"Not one of these days, sometime later," said the professor, clearly hinting that the conversation was over.

"May I write you?" asked the man.

"Do it," the professor replied impatiently as he walked into the boathouse to get the oars for his boat.

"Who was that man?" asked Jack as they walked over to the pontoon.

"Jack," said the professor, "that was Velikovsky. Remember the book I gave you? He is the man that brought the planets into disorder. Trying to explain catastrophes from the past with cosmic events. I think it is a bit crazy, but I thought you might like it. Anyway, the science is maybe not so good, and he always wants to talk to me about it."

"OK professor," Jack replied. He remembered

getting the book, but had not got around to reading it. "But now we have some sailing to do. Where is your boat?"

"My little Tinef is just over there," said the professor, pointing towards a small wooden boat.

"Die Dukas named it," he added with a chuckle. "It is a Jewish word meaning a piece of junk."

Jack laughed. It was a good name. The boat did not look like much. He helped the professor into it before undoing the moorings. They floated slowly away from the pontoon, rocking side to side as the waves clucked against the hull of the boat. Jack got up to set the sail, leaving the professor in charge of the steering. There was a gentle breeze, not enough to fill the sails, but after some trimming they set off at a decent pace.

The two of them were quiet for a long time. It was a nice day to be out on the lake and they were enjoying themselves. There was no need for talking.

*

"You know, Jack," the professor said a while later, noticeably more relaxed. "I am not sure if Die Dukas warned you, but I do not know how to swim."

"Uhm, I've heard," Jack replied. "But how can that be? You go sailing a lot."

"Yes," agreed the professor, "and I have had quite a few accidents."

Some of the familiar glint had returned to his

eyes.

He explained how he learned to sail as a young student. How it became a passion, something completely different from his academic life. A passion unburdened by technical knowledge. He admitted that he sailed absentmindedly, in a dreamy kind of fashion, mainly following his instincts. And there had been many mishaps. His mast fell down regularly. He had to be towed home. He capsized and almost drowned himself. Numerous times he had to be rescued by other boats.

"I remember the summers I spent on the coast," he smiled at Jack. "Sailing around the secluded coves was more than relaxing. I even had a compass that glowed in the dark, like a serious seafarer. I may not be talented in this art, but I always managed to get myself off the sandbanks on which I became lodged."

As he was speaking, the sadness returned to his face.

"And now it is over," he complained. "The doctors have taken it all away. It was bad enough telling me not to smoke."

"That they can never check, at least," he said, admitting to hiding a tiny pipe and tobacco in his desk at the Institute. "But now they want the boat, as well."

Helen was right, Jack decided. The old man was definitely sulking.

He checked his watch to see if it was time to turn back. It was a glorious day and he did not want to deny the professor the pleasure, but he had promised Helen they would be back well in time for supper. She had given him strict instructions not to overdo it.

As they turned around, the professor gave Jack a serious look.

"It is not just the sailing," he said thoughtfully. "I always brought a notebook. I used to love it when the sea was calm and quiet. I did not mind if the wind died and the sails drooped. It was like nature telling me it was time for a little think."

"I will miss the waves," he continued. "I have spent so much of my life thinking about waves. When I was young I was fascinated by waves of light. Later I was trying to understand space and time. How they are warped and why this leads to gravity. To my surprise, I found that there could be waves of gravity, as well. Since nothing can move faster than light any change in gravity moves as waves across the cosmos."

"Although," he sighed, "these waves are tricky. I still do not know if they are real. If they are, then we are bathing in spacetime ripples all the time. Stretching and squeezing us ever so slightly. It is a strange idea... Maybe one day we will find out."

The old man looked out over lake, seemingly lost in thought. The rippling water reflected the bright sunlight, warping and twisting it into an unpredictable pattern. The gentle waves lapped against the side of the little boat.

"How am I going to escape now?" he asked. "Everyone wants a little piece of me, and I will not even be able to hide on the lake. There will be no peace. I will be trapped in the house like an animal in the zoo."

Jack knew this was not an exaggeration. He had seen how people behaved around the professor. As if his fame would somehow rub off on them. Jack had often been impressed by the old man's patience, but he had seen the toll it was taking.

"You know, Jack," the professor spoke gently as the boathouse came back into view, "when it has once been given you to do something rather reasonable, forever afterwards your life is a little strange. There have been so many things written about me, but I do not understand why. My life is a simple thing that would interest no one. It is a known fact that I was born, and that is all that is necessary. But because of this peculiar popularity I have acquired, anything I do is likely to develop into a ridiculous comedy."

He was quiet for a few moments, before adding, "With me every peep becomes a trumpet solo."

CHAPTER 7

A theory of nothing

THERE'S A BEGINNING and an end to everything. But the end is often the beginning of something else. Things change all the time, and this can be exciting, something new.

*

Jack's life changed dramatically when he got to Berkeley. For the first time, he lived away from home, and life on the west coast was very different from Princeton, New Jersey. From the hills and winding streets of San Francisco, with the Golden Gate bridge and the Alcatraz prison island, to the rugged coastline washed by the Pacific surf. Jack had never seen anything like it.

His father had taken him to New York a couple of times. They had been to Times Square and caught a baseball game. He fondly remembered getting his beloved Yankees cap. But they had spent most of the summer holidays on his mother's family farm in the Midwest. A world of golden cornfields stretching for miles and miles, all the way to the horizon. Flat, so very flat. The west coast was something else entirely.

Life as a student was exhilarating. He went to lectures, learned new things and met new people. Different people. People with all sorts of interests.

From politics to history and philosophy. He made new friends. One new friend, in particular.

*

It took Katherine months of careful persuasion to convince Jack to bring her back home to meet his parents. He tried to wrangle out of it several times, but as the Easter break approached he gave in.

He need not have worried. His parents took to the girl immediately. It was almost as if they were Kate's family and he was the visitor. An odd experience, but Jack was pleased.

Of course, there were uncomfortable moments. Especially when his mother got a little bit too carried away and started talking about weddings and grandchildren. Luckily, Kate did not seem to mind. She just laughed and moved on to talk about something else. Jack was amazed. If the roles had been reversed, he would have run away.

The discussion of his new career plan had been more difficult. His mother could not accept that he was giving up the prospects of a secure job in business to follow what she called a whim. His father was more understanding, but did his best to stay out of the argument.

"Science writer?" Jack's mother asked pleadingly. "Is that even a job?"

"Well," Jack tried to defend himself, "it may not be easy to make a lot of money, but..."

"See! That's exactly what I mean. How are you going to make a living if there's no money? How are you going to support a family?"

Jack blushed.

"But..." he implored, "it's what I want to do. And you always said one should follow the heart."

After a brief silence, Jack's mother provided what she thought would be the killer blow.

"Of course, but not when the heart is so obviously wrong."

Kate had been quiet, but now she decided it was time to stand up for Jack.

"He's a wonderful writer," she commented. "He wrote a great piece on Einstein's theories of space and time for the student paper. It won an award."

This made Jack feel a lot better. For a moment he considered admitting that the words in the article had not really been his own. They were based on his many discussions with the professor. But he could feel the balance of the argument beginning to shift in his favour so he kept this to himself.

"And he's been asked to freelance for a couple of national magazines," Kate finished. "So there will at least be some money..."

At that moment, Jack's mother realised that Kate was a determined young woman. She did not need to worry about her little boy. He would be fine.

*

Jack brought Kate to the house on Mercer Street a couple of days into the visit. He had not bothered to call to check that the professor was in. He simply assumed the old man's health had not improved since the last time.

"Jack, how nice to see you. What a surprise!" Helen exclaimed as soon as she opened the door. "And who is this?"

"Helen, this is Kate."

"Oh, I'm so pleased to meet you. Do come in. But I'm afraid the professor's not here."

"Really?" asked Jack, "What..."

"He's working," said Helen. "He's gone to the Institute. He works as hard as ever."

Smiling at Kate, she explained.

"The professor keeps very simple routines. He gets up in the morning, has his breakfast and reads the newspapers. At about ten-thirty he goes off to the Institute. He stays there until one o'clock. Then he walks home. After lunch he often goes to bed for a few hours. Then we work on his letters. You should see the letters..."

She sighed.

"I throw most of the mail in the wastepaper basket but there are still sacks of letters. Letters from all over the world. All sorts of people. The famous, the curious, the crazies... Everyone writes to him. I hate the filing of all these letters, especially since I have so little space. As you can see, I have filing cabinets even in the hallway. And there are books everywhere. So many crates in the basement. I have often wished that Gutenberg had never lived!"

"Anyway, you don't want to listen to my complaints. Come in and sit down. I'll make a cup of tea," she continued in a calmer tone of voice. "Jack, maybe you could walk over to the Institute to pick up the professor. That would be such a great surprise."

"I'll take care of your friend," she added with a wink.

*

Jack followed the familiar route along Mercer Street until he reached the well-shadowed valley that would bring him to the Institute for Advanced Study. He decided to walk down Battle Road where his parent's house nestled next to mansions and well-kept villas. After a final turn he could see the Institute, built in the nineteenth-century style, in red brick, with a cupola and spire, standing out across the field.

This was one of the most exclusive academies in the world. Jack smiled as he remembered Oppenheimer calling it an "intellectual hotel", a "place for thinkers to rest, recover and refresh themselves before continuing on their way". It was a good description.

Jack walked through the impressive entrance and found his way to the professor's office. The door was half open. He knocked gently on the doorframe and entered. The office reminded him of the first time, many years earlier, that he visited the professor's study in the house on Mercer Street. There were stacks of papers covering every surface. Complicated mathematical expressions were scribbled all over a large blackboard on the wall. The professor was sitting at his desk. His feet were wrapped in a blanket and an electric heater was burning close by, even though it was not cold in the room. There was a glass of water on the desk in front of him.

Jack had not seen the professor since the end of the summer. He was struck by how much older his friend looked. His face was rounder, his composure mellower. His entire appearance was tired.

The professor did not hear Jack come in. He continued reading.

Jack cleared his voice to make the professor aware of his presence.

The old man looked up above the rim of his reading glasses. For a moment, he hesitated, then his face split into a warm grin.

"Jack!" he blurted out. "Where on earth did you come from?"

"Helen sent me," Jack replied. "She thought

you'd like company on the walk home."

"Ach," came the familiar exclamation, "Die Dukas never fails. How are you, my boy?"

"I'm very well, thanks. Just home for Easter. And I brought a surprise."

"A surprise? Really? How interesting."

"You'll find out when we get back to the house," said Jack. "But how are you, professor?"

"Me? Well, as in my youth, I sit here endlessly and think and calculate, hoping to unearth deep secrets."

He paused for a moment.

"But I am an old man. The mind is slow. I struggle with the same problems as ten years ago. I succeed in small matters but the real goal remains out of reach. It is hard and yet rewarding. Hard because the goal is beyond my powers, but rewarding because it makes me immune to the distractions of everyday life."

Jack was astonished. Here was one of the world's most famous scientists, the man that single-handedly changed the way people viewed space and time, making his work sound as if it were a battle, and he was on the losing side.

As he waited for the professor to get ready to leave, Jack took a look around the office. The desk was covered in papers with calculations in the professor's scrawly hand. The overall impression was of something unfinished. Jack began to see why the professor was deflated.

They slowly made their way back to the professor's house. Jack prodded the old man to find out what he was working on. Why he was stuck.

"Many years ago," the professor explained, "I set out to find a theory that connects everything. All phenomena from the atoms to the universe."

"You mean, like comparing the way the electron orbits the atomic nucleus to how the planets circle the sun?" Jack asked.

"Well, not quite. You could perhaps say that I am not interested so much in the particles as in the space between them. You remember how I told you that electricity and magnetism are different parts of what we call electromagnetism. I am trying to come up with a model that combines this theory with gravity."

"That sounds awfully complicated. Are you sure it will work out?"

"There was a time," replied the professor, "when I was convinced about it. Other people were too. There was excitement. But one by one the others lost faith and started working on other things. Now it is just me. And I am no longer convinced, either."

"But why do you carry on if you don't believe in it?"

"It is not so simple. There is a passion for understanding, just as there is a passion for music.

This passion is common in children, but gets lost in most people later on. Without this passion, there would be neither mathematics nor science. Time and again the passion for understanding has led to the illusion that man is able to comprehend the world by pure thought, without experiments."

"I have been lucky," he added. "Fate allowed me to find a couple of nice ideas after years of feverish labour. Ever since then I have been trying but not been quite so fortunate. Still, thinking about the mysteries of the universe makes one feel miraculously great and important. Like a mole in his self-dug hole."

Jack could not think of anything to say. The professor was talking about things that were far from the young man's experience.

They walked on in silence.

As they turned the corner to Mercer Street, the professor stopped. He gave Jack a serious look and said, "You are young still. You have a lot to experience, a lot to learn. Perhaps you imagine that I look back on my life's work with calm satisfaction. From up close it looks quite different. There is not a single idea I am convinced will last, and I am not sure I am on the right track."

"I have spent many years searching for a theory of everything," he finished. "My time is running out and all I seem to have is a theory of nothing."

When they arrived at the professor's house,

Helen and Kate were sitting on the veranda. They were laughing. Jack led the professor through the door to the back garden.

"Oh, another visitor!" he reacted when he noticed that Helen was not alone. Then the penny dropped.

"Jack! Is she? Are you?"

"Yes, we are," said Kate, yet again saving Jack from an awkward moment.

"What a nice surprise," beamed the professor.

*

It was late in the afternoon when Jack and Kate finally got up to leave. Helen had reminded the professor that he had better have a lie down. He was supposed to be interviewed on television that evening and it would be sensible to have a rest.

"Ach," the professor complained. "You are right. I must not give the appearance of a tired old man."

He winked at Jack.

"Apparently," he said, "I will be seen and heard by some sixty million people."

He laughed his characteristic booming laugh.

"You see, I still might become famous, after all!"

Helen and the professor followed the young couple out onto the porch. As they said goodbye, the old man looked up into the sky. He turned to Jack.

"What do you think," he said, "does the moon exist only if we look at it?"

It had happened so many times and yet Jack was totally unprepared. A strange question, seemingly out of nowhere. He didn't know what to say so he just smiled.

CHAPTER 8
The letter

THE LETTER ARRIVED a week after they returned to the west coast. As soon as Jack recognised Helen's handwriting, he knew what the message would be. It had been impossible to avoid the news.

Dear Jack,

as you no doubt have already heard, the professor left us yesterday. He was in good spirits until the end, but there was a lot of pain. He had been complaining about pain in his stomach and back for a long time, and finally his body gave up. The doctors wanted to operate, but he stubbornly refused. He told me his time had come and he was ready to go. Following his final will, we have already held a small ceremony, scattering his ashes in a secret place. He made it very clear that he did not want a grave that people could make into some kind of shrine. Neither will the house turn into a museum. As you know, he was a very private man. He did not enjoy all the attention. I am sorry to bring you the sad news this way, but I hope to see you next

time you are back in Princeton.
Yours,
Helen

*

Jack travelled back soon after receiving the letter. There was obviously nothing he could do, but he was upset that he had not had the chance to say a proper goodbye to his old friend.

They were sitting on the front steps to the house on Mercer Street.

"What are you going to do now?" he asked.

Helen stared into the distance. At first he could not tell if she had heard him. Then she shook her head gently, as if to clear it from unwanted thoughts, and replied, "Once the personal things have been sorted out, there will still be the papers. There will be years' of work getting them organised. You don't have to worry about me. I'll be busy."

It was the first time Jack had lost someone close. Both his grandfathers had died before he was born. His father's mother passed away when Jack was too small to understand. He could not remember what it had felt like. All he remembered was being woken up by the telephone in the middle of the night and his parents bundling him into the cold car for the drive to the hospital. He could not remember anything but a vague sense of loss.

This time it was very different. There was so

much he would have liked to say, and suddenly it was too late.

"I feel kind of empty inside," Jack said quietly, trying to figure out his emotions. "It's like... you know how the planets circle around the sun... all of a sudden the sun isn't there anymore. The magic is gone."

There was a long silence. The two of them just sat there watching the occasional car drive down the street. The world seemed generally quiet, at ease with itself. There was no need for talking. Words could not express what they were thinking, but there was comfort in the company.

After a long time, Helen got up and said, "I almost forgot. He left you something. A keepsake. I'll go and get it."

She disappeared into the house. When she returned, she was carrying an odd-looking stick.

"What is it?" Jack wondered.

"It was a present for his birthday. He told me you would have to work it out for yourself."

The memory brought a faint smile to her face.

Jack held up the long stick. He swung it around slowly and turned it over to have a closer look. At one end there was a small cup. Hiding in the cup was a little ball. The ball was attached to a spring. It was a most peculiar contraption.

"I wonder," Jack mumbled as he tried to figure out what the thing was good for. It had to have some meaning. Otherwise the professor would not have left it for him. He tried to pull the small ball out of the cup. He could feel the spring trying to draw the ball back, but it was not quite strong enough. He left the ball dangling outside the cup. He was thinking back, trying to remember the many things the professor had told him. Then, all of a sudden, he got it.

"I know," he exclaimed, jumping to his feet. He held the stick up above his head. Then he dropped it. As it fell, the spring pulled the ball into the cup.

Jack couldn't help laughing.

"That's great!" he said enthusiastically. "Helen, did you see that?"

She gave him an amused look.

"It's his principle of equality," Jack explained. "When you hold the stick up, the spring is not strong enough to pull the ball back into the cup. Gravity wins."

He repeated the experiment. The same thing happened.

"But when you drop it, the ball can no longer feel gravity so the spring wins," he added. "Amazing. It's where everything started. Acceleration and gravity are equal. Warped space and time. Everything. What a fantastic gift!"

*

They walked over to the Institute after lunch.

"Oppenheimer has returned from a trip. He called to tell me he's planning to say a few words today," Helen explained. "I thought it would be nice to be there."

They arrived just before tea-time, the one time of day when the deep thinkers all come out of their offices. As they entered the big room, Oppenheimer gave Helen a nod.

Jack recognised the gaunt man, with the close cropped hair and the icy blue eyes. Despite the turmoil he'd gone through, Oppenheimer was still the Director of the Institute. *The manager of the hotel*, thought Jack.

"My dear friends," Oppenheimer stood up and, as soon as the room was quiet, started speaking. "As you all know, the fourth dimension, time, finally managed to overtake Dr. Einstein a few days ago. Many of you may not have had the chance to get to know him personally. You may perhaps be excused for viewing him as more of an ancient landmark than a beacon pointing the way to the future. But there is no doubt that he was a great man. His contributions are beyond assessment in our day. Only future generations will be competent to grasp their full significance.

By the sheer force of his personality, by his naturalness, his simplicity, his humour, his mastery of his subject, and the indefinable aura of greatness that no modesty could hide, he captivated his audiences all around the world. He was not simply a great thinker, he fought hard to defend the right to search for truth and to teach what one holds to be true. He will be sorely missed, but for years to come he will continue to be a source of inspiration to those who use their minds and imaginations in the pursuit of science.

In his absence, the rest of us must take up the challenge. We must move through time and space without the aid of his gigantic mind. I hope you will join me in a few moments of silence and contemplation in memory of our departed friend."

Oppenheimer remained standing, leading the silent tribute. People bowed their heads, moved by the Director's heartfelt words. Jack fought hard to control his emotions, but the difficult moment passed. Oppenheimer sat down and started chatting with his colleagues, almost as if nothing out of the ordinary had happened.

Jack managed to slip away without anyone noticing. He walked slowly down the empty corridor until he found himself outside the professor's office. He opened the door and stepped in. Nothing seemed to have changed. The desk was covered with unfinished equations and unanswered questions.

The professor had left behind a jumble of mail, books and technical journals. On top of the pile lay his beloved pipe. Jack noticed a stack of letters from all over the world, addressed in a number of languages. There were pads of paper filled with foot-long equations.

The empty chair looked as if the man who usually sat in it had merely stepped away, perhaps to gaze reflectively at the meadow that rolled past the Institute. But he was not coming back. The chair would never again be filled.

Jack caught sight of the blackboard, covered with chalk hieroglyphs. The mathematical expressions seemed to mock him. He found himself staring at them blankly. Then the chalk marks went blurry and he had to close his eyes.

CHAPTER 9

The nature of time
1964 PRINCETON, NEW JERSEY

LIFE HAS A certain rhythm. There are ebbs and flows, ups and downs. Things change with time, and time also changes with age.

As you get older, time seems to speed up. You run out of time, all the time. There never seems to be enough of it to do all the things that need to be done. You hardly notice years go by. The time between birthdays seems shorter and shorter. But maybe this is not because time is speeding up. Maybe it has more to do with how you remember things. Or perhaps all the things you forget.

When you're young, every day brings a new experience. Something exciting. When you get older, you've seen many everyday things before, so you forget about them. They become routine and the time they take up doesn't count, but there are still new experiences. Things you will never forget.

*

Jack would never forget the night the twins were born. It had nothing to do with the unusual snowfall or the mad rush to the hospital in the middle of the night. It had little to do with the pacing up and down the corridor and the long, nervous wait. It was all about that first moment when he held the two tiny bundles in his arms. The moment his heart grew at least two sizes. The moment his life changed. Soon he could no longer remember what life had been like before they became a family.

It was a frantic time.

It was always going to be a challenge to bring up twins at the same time as carving out careers, but Jack and Kate had no choice. It had to be done, and they had to manage. Somehow. It did not help that they had chosen to live so far away from their families. They could not simply drop the kids off with the grandparents whenever they needed a break. There was no one to call on at short notice. This made life a juggling act where things were often up in the air until the last possible moment.

They had no time of their own, could hardly remember what it had been like when it had been just the two of them, but they would not change this for the world.

*

"Danny, don't pull your sister's hair!"

As Jack reached out to grab the boy, the toddler's arm shot out and knocked over a cup of coffee. The black liquid splashed all over the notes for the article Jack was working on.

"Oh no, not again," he grunted.

"Jack, what on earth are you doing to the children?" his mother asked as she entered the kitchen.

"I'm just..." he started, but stopped himself when he realised how unreasonable it had been to try to get some work done while giving the energetic twins breakfast. It had seemed a good idea to let Kate sleep, because she needed a rest after the previous day's travel. The work part had not been such a good plan. The children demanded his full attention.

"Come on, you two," his mother saved him. "Let's go and have our breakfast outside. Let daddy get on with his work."

"Thanks mom," Jack said gratefully.

He wiped his notes clean. The damage was not too bad and anyway they were just rough ideas for something he was planning to write. He had been thinking about it for a while and when he learned that professor Wheeler was going to give a series of public lectures he made his decision. It was time to write that article about time.

*

Jack walked over to the Institute with Helen, who had received a personal invitation to the talk. It had been many years, but she was still busy organising the professor's papers. The task seemed endless. Jack asked if she ever got fed up with it.

"I guess there are times when I get a little bit bored," she admitted, "but that probably happens no matter what you do."

"Actually," she added, "I'm thinking of writing a book."

Jack was astonished.

"Really? What kind of book?"

"Well," she explained, "throughout the many years I worked with the professor, I kept a secret box. My little box of snippets. Whenever I came across something he'd said that I found charming or striking, I used to type up my own copy and put it in the box. It's great fun looking back on these snippets now. I have quite a collection. I thought it would be nice to put them together. It would be a book about what he was really like. As a human being."

Jack thought for a moment.

"You know," he said, "I think that's a really good idea. There's been so much written, but mostly by people who didn't know him. And you knew him better than anyone else. I would certainly want to read it."

They walked on quietly, not wanting to disturb each other's memories.

As they crossed the grassy field in front of the Institute, Helen broke the silence. "Do you remember John?" she asked.

"You mean Wheeler?" Jack replied. "Of course, I do. I remember how he used to bring the students of his gravity class to the professor's house for tea. They asked all sorts of crazy questions."

"They certainly did," Helen chuckled. "All those years ago. John still visits every now and then, you know. Tells me about the new things they've discovered. The professor would never have imagined. John and his students have done so much to explain how things work."

"They sure have," Jack agreed.

"And you know how they describe him?" he asked a moment later. "John Archibald Wheeler, possibly the greatest physicist you never heard of."

They both laughed.

*

The lecture hall was nearly full when they arrived, but they found two seats at the back of the room. The atmosphere was one of excitement and anticipation. Wheeler was famous for his high-energy lectures and his entertaining explanations of complicated phenomena.

"Good evening, Ladies and Gentlemen," he started. "Tonight I want to talk to you about time. The nature of time.

What is time?

Why do we need it?

Is it simply there to avoid everything happening at once?

There are so many questions and I don't have many answers. But years of experience have taught me that you never really understand anything until you try to explain it. So I'm quite hopeful I might learn something before the end of the evening."

There was scattered laughter across the room.

Wheeler started writing rapidly on the blackboard using both hands, regularly twirling around to make eye contact with the audience. Soon the blackboard was filled with colourful drawings and a splatter of mathematical expressions.

"Human beings have always tried to understand the world," he explained. "Back in the days of the great explorers, like Columbus, most people thought the earth was flat. They believed the Atlantic Ocean was filled with monsters large enough to destroy their ships. At the edge there were fearsome waterfalls over which brave sailors would plunge to their destruction. Looking back further, we have the myth that the world is held up by elephants, ultimately resting on the back of a giant turtle."

He made a rough sketch of a turtle.

"Of course, these days we know these things are not true. But our modern view of the world is almost as peculiar. We may think we're more advanced, and perhaps we are, but ultimately there's still much we don't understand about the universe. Among the many problems, time may be the most mysterious.

Why does time flow from the past to the future?

Why not the other way? Some scientists argue this has to do with the laws of thermodynamics, the relationship between heat and other forms of energy. Everything moves from order to chaos. You know, without doubt, that a glass you drop will shatter and splash its contents all over the floor. It's never the other way around. It never happens that you drop a broken glass and it lands put back together and full to the brim."

"Unfortunately," he paused briefly for effect, "our theories do not tell us why this has to be the case. It's a tough nut to crack."

He took a sip from a glass of water before continuing.

"But one thing we have learned is that time is flexible. We know this because of the work of one man. Albert Einstein. Without him, the theory of relativity may have laid long undiscovered.

When I was a much younger man, I used to look at Einstein as firmly in the past. An old man following his own track in a direction pointing away from the main stream of physics. I now find it strange that I could so cavalierly dismiss the greatest thinker of the century. It was later, when I moved to this beautiful little town, that relativity became my passion. I was lucky to become friends with Einstein and had some wonderful conversations with him in his house on Mercer Street.

The impact that Einstein's ideas about space and time have had on science cannot be exaggerated. He showed that time slows down on a moving clock. For a long time, it seemed as if the effect would be too small for us to test it, but this has changed. We now have very precise atomic clocks that track the tiniest variations in the speed of time. We can prove Einstein right.

However, there is another effect in Einstein's theory. Because gravity warps space and time, a clock at higher altitude should tick faster than one on the ground. You may not be able to tell the difference, but your head is actually a tiny bit older than your feet.

For me, this is the most exciting part of the story. Space, time and gravity are all part of the same structure."

He took another sip of water to let the message sink in.

"Now let us take this idea of time slowing down to the extreme. If gravity slows time down, can it make a clock stop altogether? Yes, we believe it can.

In Einstein's universe there are exotic objects that are formed when matter, collecting in an ever-shrinking space, creates so much pressure that it can no longer support itself and starts falling into itself. If this collapse continues, you eventually reach a point where all our theories of predicting what happens fall down.

We are confronted with the puzzle of what happens next when there is no next.

This may sound like absolute madness, but we are beginning to find evidence that these weird objects exist in the cosmos. We call them black holes.

We have no means of observing what happens if anything falls into such an object.

Imagine a space explorer riding along with the infalling matter. If he were to send a radio message from the collapsing mass, the space through which the message travels would crunch against the signal and prevent it from getting out. This poses many fascinating problems for us to think about."

"Space that looks so empty and free of structure is in fact full of structure," he continued after a brief pause.

"We're only beginning to understand the many possible complexities. One idea, which may ultimately teach us something deep about the nature of time, is that of wormholes, tunnels that connect different points in space together."

"It would be as if I could put my finger through a hole here..." Wheeler said as he walked across to the other end of the stage, "... and see it emerge in a hole way over here."

There was a confused murmur from the audience.

"These wormholes may allow Alice in Wonderland-like loops in time. If they were to exist one could, in principle, live one's life over and over again. I'm not sure I like that idea, but I guess one would eventually learn not to repeat the same old mistakes."

He took a deep breath before finishing the lecture with a flourish.

"We've been on a journey from the ancient theories of the world to the current thinking about space and time. Where are we going? What about the future?

Will some of the ideas that are science fiction today become science fact tomorrow? Yes, I am sure they will.

What about the nature of time, will we ever understand it? Yes, I am convinced we will. One day.

Will we even be able to conquer time, to travel freely in any direction, to build a time machine? Personally I am doubtful.

After all, if this were possible, why are there no tourists from the future here to gawk at us?"

Wheeler stopped and smiled.

"Thank you for listening and good evening," he finished.

The audience applauded enthusiastically.

CHAPTER 10

A black hole is not a hole
1972 PRINCETON, NEW JERSEY

NEIL ARMSTRONG TOOK a small step onto the surface of the moon.

Jack and the kids were glued to the small black and white television. The picture was fuzzy and the sound crackled. The children were too small to really understand, but Jack knew this was one of those moments he would never forget. They had finally arrived in the future. The past was history.

*

John Wheeler was sitting at his desk, skimming through the Daily Princetonian's report on his latest lecture. He frowned as he recognised some of his colourful explanations of space and time.

He put the thin newspaper down, grabbed a notepad and picked up his trusted fountain pen. He was just about to jot down some thoughts on how he could explain things better when he was interrupted by a knock on the door. It was the department secretary.

"Excuse me, professor," she said. "There's a reporter here to see you."

"But," he gave her a surprised look, "I don't remember any appointments today."

"No," she confirmed, "the young man said he

just dropped by on the off chance you'd have a few moments to talk to him. Apparently he would like to check a few things you said in your talk the other night."

Wheeler did not like being ambushed by reporters, but he was not particularly busy this morning.

"OK, then," he decided, "I'll spare him a few minutes."

*

"It's you!" he exclaimed the moment the secretary ushered Jack through the door. "I know you. It's... Jack, right? We met at Einstein's house. Years ago."

"That's right, professor," Jack confirmed. "I'm surprised you remember."

"I remember it well," said Wheeler, "it must have been a year or so before he died. I had started teaching relativity and he was kind enough to invite me and the students over to the house for discussions. We sat at the table, Helen brought tea, and the students asked all sorts of questions. I remember you were there a couple of times. It's great to see you. What are you up to these days?"

"I'm a journalist. Freelance. I actually came to see if I could talk to you about an article I'm writing for the National Geographic. It is supposed to be about space and time, but I don't want to take up too much of your time."

"Time?" Wheeler pretended to check his watch.

"Don't you know it depends on how you look at it? The way I see it... we have all the time in the world."

He gave Jack a beaming smile and gestured towards one of the chairs on the other side of the desk.

"Thanks," said Jack. He opened his notebook and gathered his thoughts. "Can I ask about that course you taught back in those days? How did you get interested in gravity in the first place?"

"That's a great question," Wheeler replied. "It's funny to look back on it now. At that time, Einstein's theory did not, as far as I can tell, attract much interest. Perhaps people thought it was all done and dusted. There were no new experiments and no exciting problems to work on. People just thought of it as messy mathematics."

"How things have changed," he added thoughtfully. "Anyway, I enjoyed the teaching. I was learning as I went along. It took some time to get the idea of bent space through my thick head. What it really means. That there's a spacetime of events like grains of sand on a sheet of sandpaper, laid out next to one another. And then I started to see that there were many interesting questions. I was particularly excited by the unsolved aspects of gravitational collapse."

"I'm not sure I understand what you mean by collapse," Jack admitted. "Can you explain that part?"

"Sure. It's a very important process, where gravity keeps pulling things together until something intervenes. Maybe that's not much clearer. Uhm... let's see. It's through gravitational collapse that stars are born, that clusters of stars are formed and entire galaxies are created. That part is not very mysterious.

Now, I'm only ever happy when swimming in a mystery, and I kept asking where was the mystery in this gravity business. There was a question of the collapse of a star. The mathematics seemed to say that a heavy star would just keep falling in on itself, but this was hardly the kind of mystery I was looking for.

Besides, I think everyone agreed that the conclusion was wrong. Einstein certainly didn't believe in it. People just didn't think the continued collapse theory gave an acceptable answer to the fate of a star. I remember having a fight with Oppenheimer about it. Then I realised we were dealing with a new phenomenon, with a mysterious nature of its own, that was waiting to be unraveled."

Jack scribbled frantically in his notebook, trying to keep up with the flow of information. He had many questions, but Wheeler was like an unstoppable freight train.

"Look at it this way," he continued enthusiastically, grabbed his fountain pen and drew a quick sketch on a piece of paper. "Imagine a star that is large enough. Much heavier than the sun. This star will eventually die in a grand cataclysm. As the nuclear fires begin to burn out, the stellar gases, no longer supported by heat and radiation, begin to fall towards the star's core. Moving at tremendous speed, they crush together, forming a ball only two or three miles across, so dense that each cubic inch of material weighs trillions of tons. The gravity of this tiny sphere is so strong that no radiation, not even light, can escape. It has become totally invisible. It has become what we call a black hole."

"This sounds... uhm... too fantastic to be true," Jack commented with a frown.

"It may sound absolutely crazy," Wheeler agreed, "but astronomers are finding evidence that these objects are out there. We're learning that the universe is full of strange phenomena, and some are just a bit stranger than strange. Take the quasars, for example. We know they are distant galaxies, but the energy you need to power them is enormous. And they vary rapidly. This tells us they have to be small. A black hole would be small enough, and the strong gravity would be able to release all the energy you need."

"Although, we don't know how they do it," he admitted. "And we don't understand how a black hole can grow that big. Some of these objects are absolute beasts, weighing more than a billion suns."

Jack's jaw dropped slightly. He let out a faint whistle. "Phew," he said. "That's seriously big."

"Absolutely", Wheeler agreed, "but the universe is an amazing place. We've come a long way in less than ten years. Astronomers are opening up an entirely new cosmos, only vaguely dreamed of even in Einstein's day. The night sky is no longer this serene place where stars glide gently along their way and galaxies evolve over eons.

Reality is quite different. Many stars end their lives in spectacular explosions that shine brighter than an entire galaxy. We see x-rays emitted as black holes feed off their partners in double systems. There are spinning stars that act as radio lighthouses. We call them pulsars. The universe is a violent place. In fact, the whole thing started in an immense explosion. The big bang. We know that galaxies fly apart as the universe expands.

Once you start to see all this, the idea that a dead star may collapse into a heap so dense that light can't escape from it doesn't seem quite so strange. A black hole is often the simplest explanation for what we're seeing. You could perhaps say that black holes have jumped off the blackboard and into reality."

Wheeler gave Jack a mischievous smile.

"Of course, there are still mysteries. According

to the mathematics, the star would collapse forever while spacetime wrapped around it like a dark cloak. At the centre, space would be infinitely curved and matter infinitely dense. It's like an entire star being squeezed through a knothole.

The black holes teach us that space can be crumpled like a piece of paper into a tiny dot, that time can be extinguished like a blown-out flame, and that the laws of physics we regard as sacred are anything but. It's thrilling that the world is such a mysterious place."

"I have to confess..." Jack started, "It's a bit overwhelming. I'm struggling to picture these things. What would they look like?"

"What would it look like?" Wheeler thought for a moment before replying.

"Like... like... It had the biggest head you ever saw... A huge great enormous thing, like... like nothing. A huge big... well, like a... I don't know... like an enormous big nothing..."

Jack couldn't help laughing.

"I'm sorry, professor, you wouldn't even fool my kids with that. You stole it from Winnie the Pooh."

"Ah yes, of course, I'm sorry," Wheeler said, even though he clearly wasn't. "The problem is I don't really know. The matter that formed the black hole has long since disappeared, like Alice in Wonderland's Cheshire cat, leaving behind only the disembodied grin of its gravity. From far away, that gravity has the same effect as it did when its matter existed. But closer in, the gravitational force soars, becoming so great that it prevents the escape of light. It's difficult to describe what this would look like."

"Is that why it's called a black hole?" Jack asked. "The matter that used to be there seems to vanish from the universe."

"Hmm, yes," Wheeler confirmed, "although a black hole is not a hole. We're not talking about a rip in the fabric of space and time. You could still imagine someone falling in, an astronaut perhaps. Once inside, he would never get back out, or communicate whatever he is seeing to the outside. He might solve the mystery of what's going on inside, but we would never find out. It might seem a bit pointless, but who is to deny someone the right to their personal pursuit of knowledge?"

"So, what you're telling me is that the name doesn't really describe these things? They're black, but not holes?"

"Actually, they might not be quite black either."

"But isn't that terribly confusing."

"Sure, but the old names were even worse. The Russians used to talk about frozen stars, because someone watching from far away would never actually see anything falling through the surface. Gravity is so strong that time would freeze. It's not a great description because a clock that falls in doesn't actually stop ticking."

"Is that the relativity part of it?" Jack tried to connect the conversation with the many things Einstein had told him.

"Correct. In the West we used to call them completely collapsed objects. That's a terrible name, too. Once you use that long-winded phrase five or six times, you have to look for an alternative. If you want people to believe in your idea, it helps if you give it a good name."

"So where did the name black hole come from?"

"I think it was an act of desperation. It might even have started as a joke. I had been searching for the right name for months, mulling it over in bed, in the bathtub, in my car, whenever I had quiet moments. Suddenly this name seemed just right. And it stuck."

Jack jotted some things down in his notebook. Looking through the notes, he decided he had more than enough to write his article, but he was still curious.

"We should probably finish. I've taken up too much of your time already. Before I go... do you mind if I ask a stupid question?"

"Not at all. In fact, no question is stupid enough not to be interesting."

"OK, then." Jack thought for a moment before continuing. "If you were to describe yourself in a few words, how would you do it?"

"Ah," Wheeler leaned back in his chair and looked up at the ceiling.

"I guess I shouldn't be too flippant," he chuckled. "But one thing is clear. I'm against logic. I like the clash of ideas. It makes you think. If there's one thing I feel more responsible for than any other, it's the perception of how everything fits together."

"The universe is a complicated puzzle and we only have a few of the pieces," he explained. "I like the really big questions. For example, is man an unimportant bit of dust on an unimportant planet in an unimportant galaxy somewhere in the vastness of space?"

"No! That can't be," he answered himself with conviction. "The need to produce life lies at the centre of the universe's whole machinery and design. Without an observer, there are no laws of physics. Why should the universe exist at all? The explanation must be so simple and so beautiful that when we see it we will all say, 'How could it have been otherwise?'"

"But we're not there yet. We still have a lot of thinking to do," he finished.

CHAPTER 11

Bedtime stories

THE RIVER OF time gently meandered downstream. Jack and Kate reached middle age. The children became teenagers.

On his 40th birthday, Kate gave Jack a classic Underwood typewriter. It was love at first sight. Everything about the machine was perfect. From the rhythmic click-clack of the keys to the cheerful bing every time he finished a line.

"It's almost as if it is writing itself," he told Kate. "It's absolutely brilliant."

"Of course it is, darling," she smiled back at him, "and I'm sure it will help you write like your heroes Kerouac and Hemingway, but the truth is... I only got it to be able to check that you are actually working. I can hear your typing all the way from the kitchen, you know."

Jack was about to protest that he wasn't likely to hide away in the office just for the sake of it, but Kate was not quite finished.

"But now I wonder if this wasn't a big mistake," she added. "I'm beginning to think you care more about Mr. Underwood than you care about your family."

"I'd better leave the two of you to it," she finished and pretended to stomp out of the room.

*

"What are you doing, Liz?" asked Kate.

The girl was sitting on the steps to the veranda at the back of the house on Mercer Street. She was watching her brother, halfway up one of the many trees in the back garden.

"I'm waiting for Dan to drop like a rotten fruit," she replied.

"Oh Dan, be careful," Kate called out as soon as she saw what the boy was up to.

"I'm OK, mom," he insisted.

"You still have to be careful, son. You know where you are. Gravity can be temperamental," Jack commented. "You won't know you're falling until you hit the ground."

"And anyway," Helen added, "you don't know what the professor may have left lying around. You don't want to fall into one of those wormholes. Who knows where you'd come out."

"Or when..." Jack filled in.

He leaned back in the chair and enjoyed the warmth of the afternoon sun. He took another sip of ice tea and reached out for a piece of the cake Helen had brought through from the kitchen.

"So, Helen," asked Kate, "are you still working on the professor's papers?"

"I'm afraid so. It seems to take forever. I think

107

we must have gone through many tens of thousands of documents. There are so many letters. I remember how we used to struggle trying to answer them all. Back in those days letters arrived in scores, from great men and humble ones. From politicians, scientists and cranks. At least the flood has stopped now."

"He seems more popular than ever," said Kate. "His picture is all over the place."

"You're right," Helen replied. "Do you know what he would say? 'You see, they are still taking pieces out of my hide.'"

Her mock German accent made everyone laugh.

"Actually," she added wistfully, "maybe we all are... I mean, here I am working my way through his papers. And I still put flowers in the study upstairs."

"I think that's fair enough," said Jack. "It would have been his 100th birthday last week. Surely, that's worth a few flowers?"

"Oh, I almost forgot," said Helen. She got up from her chair and went back into the house. A couple of minutes later she returned with an album of some sort. "I've got something to show you."

"Aunt Helen, are you starting a stamp collection?" asked Dan, who had given up climbing and joined them at the table.

"I guess you could say that," she replied. "Many

countries are making stamps to celebrate the pro-fessor's birthday. I was sent some, and I thought it would be nice to put them together. They're from all over the place."

"Can I have a look?"

Helen handed Dan the album and he started flicking through the pages.

"The USA, of course," he commented, "Italy, Germany, Ireland... and look at these! Soviet Union, Cuba, Vietnam. This is amazing."

"Oh, that's a pretty one," said Liz, looking over her brother's shoulder. "The Republic of Gabon. Where on earth is that?"

She was pointing at a colourful first-day stamp in gold and dark purple. The image had the pro-fessor playing the violin with a background sprinkled with mathematical expressions.

"Let me see," asked Jack.

The children gave him the album.

"Ah yes, that is nice," he agreed.

He checked the time on his watch and looked over at his wife.

"Kate, I think we'd better go. Don't want to be late for the play," he said. "Helen, is it still OK for Dan and Liz to stay with you?"

"Of course," Helen replied. "It will be a pleasure."

Helen had for many years been the children's favourite babysitter. The twins had pretty much adopted her into the family. They loved their Aunt Helen's crazy stories about professor Einstein, the way she emphasised his sense of humour and detachment from the world of lesser mortals.

"And you two," Kate told the twins, "had better behave yourselves. No squabbling."

"Of course not, mom," they chimed in unison.

"As I'm the oldest, you can safely leave me in charge," Dan added with a big grin on his face.

"Oldest," Liz complained. "You may be a couple of minutes older, but not one bit wiser."

"At least I'll always be older than you!" grumbled Dan.

"That's exactly what I meant," Kate interrupted the budding argument.

The twins were like two peas in a pod. Clever with sharp tongues. Most of the time they got on, but they did enjoy winding each other up. The older-wiser argument was a familiar one.

"Anyway," said Jack as they were leaving, "I think maybe Helen can tell you why the fact that Dan was born first doesn't mean he must stay the oldest."

*

"What on earth was that? What was he talking about?" Liz asked as soon as the door closed behind the parents.

"No idea," said Dan. "I'm older than you, and that's the way it's going to be. End of story."

"Helen, do you know what he was on about?"

"Hmm, yes, I believe I do," Helen replied. "He was thinking of the professor's twin paradox."

"The twin para-what?" asked Dan.

"Paradox, Dan," his sister explained. "It's a fancy word for contradiction."

"Sure, I know that, but what do the twins have to do with it?"

"Let me tell you," Helen started. "You probably remember me telling you how time is relative. The rate at which a clock ticks depends on how fast it is moving. A clock that speeds along slows down, if you see what I mean."

"I remember that," Liz confirmed, "but I don't

see what difference it makes."

"Well," Helen continued, "let's imagine we build a space rocket. It has to be one that can travel near the speed of light. Then we send one of you on a trip to the stars. Now, since the rocket would be moving at high speed, any clock on board will run slow compared to a clock left back home. So the travelling twin will age slower."

"I get it," laughed Liz. "We send Dan into space and when he gets back, he will be younger and I'll be the oldest. Great!"

"But what about the para-whatsit?" Dan insisted. "Where's the contradiction?"

"Well spotted, young man," praised Helen. "Ask yourself what happens according to the person that went on the trip."

"Uhm," the boy thought for a moment. "First the earth would travel away from him, and on the return trip it would come shooting towards him. Either way, clocks on earth would be moving relative to him..."

"... or her," he added with a glance at his sister.

"So..." he hesitated before drawing the conclusion, "this would mean the person that stayed at home had to end up the youngest!"

"Correct!" Helen confirmed. "There's your paradox."

"But it doesn't make sense," said Liz thoughtfully. "Doesn't this prove the professor's relativity theory has to be wrong?"

"You might think so," Helen agreed, "but there is another answer. I'll let you think about it while I make us something to eat."

*

"You two have been awfully quiet. Did you solve the problem?" Helen asked as soon as the food was on the table.

"Not sure," Liz admitted, "but we think the answer should have something to do with how the rocket speeds up and slows down. The acceleration must be the explanation. Somehow."

"Very clever," smiled Helen. "You're absolutely right. The relativity of the clocks did not allow for things speeding up in the first place. To work it out you actually need to use the theory that has gravity in it. Well done. The professor would have been proud of you."

*

After supper the youngsters helped Helen tidy up in the back garden. It was a pleasant evening. The air was still warm, even though there was a slight breeze. When the job was done, they sat down on the veranda to enjoy a well-deserved cold drink.

"Thanks a lot, you two," said Helen. "That was much appreciated."

"Oh come on, we did hardly anything," Dan replied.

113

"The way you look at it perhaps... but it's relative. You've been a great help. This old body is beginning to find it difficult to lift things. And don't get me started on trying to bend down or reach up high."

Helen sighed.

"That's what it's like getting old, I guess," she lamented. "Time doesn't seem to slow down even though I keep myself busy."

"So thank you very much," she finished. "I wonder how I can possibly pay you back."

"I know," said Liz, "how about a story? Like the bedtime stories you used to tell us when we were little."

"Something funny!" Dan added.

"A story?" Helen thought for a while before she made her decision. "Alright then. Here's one you used to really like."

"It must have been almost forty years ago," she started. "At that time, the professor always tried to spend his summers where there was good sailing. Several years we went to Long Island. He used to love taking his funny little boat around the many coves. I remember walking down to the water towards the end of the day. I used to stand on a rock looking out over the lake waiting to see the professor's boat come into sight so I would know it was time to put his baked potato in the oven."

She smiled at the distant memory.

"This would have been at the beginning of one of those stays. One day the professor went into the local store. The people in the store must have recognised him immediately, but they treated him just like any other customer.

After a while, the store owner asked 'Are you looking for something in particular?'

'Sundials,' the professor answered with his German accent.

Now, this shop had a large variety of items, just about everything from housewares, to fishing tackle and bait, to hardware, to toys, to appliances. But no sundials. Not for sale, anyway. But...

'I do have one in the back yard,' the shopkeeper said.

Then he led the professor, who must have been a bit bewildered by it all, out into the back yard, to show him the sundial.

'If you absolutely need one, you can have this,' the man said.

The professor took one look and began to laugh. He pointed to his feet.

'No. Sundials,' he said.

Sandals. Those, they had in the shop."

The children laughed.

"The best part of the story," Helen added, "is that the professor and the shop owner, I believe he was called Mr. Rothman, became great friends.

They used to play violin through the evenings. Often annoying the neighbours... but that's another story."

*

"I'm sorry we're late," said Kate when they returned to pick up the twins. "I hope we haven't kept you up, Helen."

"No problem," Helen replied. "It's been a pleasure."

As they drove the short distance to Jack's parents house, Liz and Dan told their parents about the evening.

"Oh, the old sundial story!" Jack exclaimed as soon as they finished. "She's told you that one before. Years ago. I remember it made quite an impression on the two of you."

Kate chuckled at the memory.

"Sure did," she said, "it was impossible to get you to bed that night. You were running around like lunatics shouting, 'Sundial! Sundial!' at the top of your tiny voices. In the end you just collapsed from exhaustion."

"Hope there won't be a repeat tonight," said Jack, trying his best not to laugh.

There was silence in the back of the car.

CHAPTER 12
Weighing the cosmos
2006 SEATTLE, WASHINGTON STATE

"WHAT ARE YOU writing, dad?" Liz asked as she entered her father's study.

"Oh," Jack reacted with a start. He straightened up from the hunched position in front of his laptop. "I'm trying to finish this infernal piece on the cosmos for Scientific American, but it's hard going. I keep getting distracted."

"Sorry," said Liz, "I didn't mean to disturb."

"It's not you. It's the memories. I'm trying to write about the present, but I keep thinking about the past. Everything's connected. I can't get the story straight."

"After a lifetime of writing, I still don't seem to be able to deal with deadlines," he sighed. "Time just runs out. Every single time."

He opened one of the desk drawers and took out an old photograph.

"And it feels like I've come full circle," he said, handing the picture over to his daughter. "I have reached the age the professor was when I first met him. I was just a boy then and I am an old man now."

"Unfortunately, my hair is less impressive," he added, stroking his bald head. "And I am not as wise."

"What is this picture?" asked Liz. "I've never seen it before."

It was a black and white photograph of a boy and a wild-haired old man on the front steps of the familiar white wood-framed house. A much younger version of her father next to Albert Einstein. The professor was wearing a pair of ridiculously furry slippers.

"I found it among your mother's papers," Jack explained. "I had no idea she had it. You know, if the photographer hadn't sent me on that errand to Mercer Street all those years ago, I would not have met the professor."

"It was a different world. The pace was slower. Nothing had to be done straight away. Deadlines were easier to deal with," he added. "With the internet, laptops and clever phones, everything you need to know is at your fingertips. All the time. But is that actually good? Call me old-fashioned, but we seem to have forgotten how to interact with the real world. Without screens. How to touch it, smell it, take time to experience and enjoy it."

"I don't think the professor would have enjoyed this modern world," he decided. "He was a gentle wizard. Needed time to think. He may have been the smartest man on earth, but he would never have guessed how things would change. There's no way he would have expected his ideas to become part of everyday life."

"I'm sure that's right, dad," Liz agreed, "but I think you are exaggerating a bit. I mean, gravity is obviously important. It is useful to know that things fall down. Never up. But other than that... Do we really need all that wonky space stuff?"

"You may not think so," her father conceded. "However..."

He pulled out a well-thumbed file from a stack of papers.

"... let me read you something. Again, I had no idea your mother had saved this old article."

He cleared his voice and started reading.

"A couple of weeks ago, two scientists, Joseph Hafele and Richard Keating, joined the passengers of an ordinary airliner from Washington to London. They carried with them one of the world's most precise atomic clocks. The clock took up two seats, while the scientists had to settle for one each.

After London, they continued eastwards around the world. Tourist class, all the way. Then they turned around and travelled westwards to get back home. When they finally landed, they compared their well-travelled clock to a twin that had been left in the laboratory. The two clocks did not agree. But the disagreement was exactly as predicted by Albert Einstein's theory of relativity. It was an amazing result."

He paused a brief moment for effect.

"The disagreement of the two clocks was not simply due to the fact that one had been moving while the other one did not. Because gravity warps space and time, a clock at higher altitude ticks faster than one on the ground. In order for the clock experiment to make sense you have to account for this effect, as well. The perfect agreement with the theory proves Einstein's lasting legacy. Space, time and gravity are all part of the same structure."

"I guess that's interesting," said Liz, "but I don't think many world travellers will be worried about an effect so small that you need an atomic clock to measure it. Hardly everyday life, is it?"

"Fair enough," Jack agreed, "but how about this one?"

He riffled through the folder and picked out another faded newspaper clip.

"Did you know? The multi-billion dollar Global Positioning System has 24 satellites orbiting the earth, each carrying an accurate atomic clock, allowing precision navigation across the world.

If we want to understand how accurate the flying clocks have to be, we first of all need to understand the speed of light. Now, you know how fast you can run, and you probably know how fast a train goes. You know you would not be able to catch the train. You are too slow. But nothing can catch light. It is the fastest thing in the universe. Light travels a million miles in a few seconds. Be-

cause light is so incredibly fast, the clock in each satellite must be accurate to better than a millionth of a second.

The satellites circle the earth twice a day, moving much faster than clocks on the ground. Albert Einstein's theory of relativity tells us that the moving clocks will slow down. However, in Einstein's curved space-time, gravity makes the clocks on the ground move slower. This makes the orbiting clocks move faster. Combining the effects, the moving clocks speed up by about 40 millionths of a second. You may think this is a tiny effect, but we ignore relativity at our peril. If we were to ignore our famous scientist's findings, navigational errors would accumulate at the rate of several miles every day. More than enough to make Satnav driving dangerous."

Jack gave his daughter a triumphant look. She could not help laughing.

"I give up," she chuckled. "You win, as usual."

"But, I am confused," she added thoughtfully. "What has this got to do with your article on the cosmos? It's obviously nice to be reminded of these old clips, but I can't see why you are spending so much time on the past."

"Ultimately, it's all about time," Jack replied. "You see, when we look out into the universe we look back in time. I just didn't look back far enough yet."

Suddenly his face was lit up by a beaming smile.

"That's it!" he exclaimed. "I've got it. I've got the angle. I know how to write this thing. You've been a great help, Liz. Always helps to talk through your problems."

Liz knew this was the signal for her to leave. Her father's mind was already elsewhere. She slipped quietly out the door.

An hour later Jack sat back in the chair. He closed his eyes for a moment. Then he scrolled to the top of the document and started reading.

Over the last half century, Albert Einstein's celebrated theory of gravity has been tested to high precision. It has passed all tests with flying colours, both in the solar system and in stronger gravity regions involving neutron stars. We have convincing evidence that black holes exist. But what about the cosmos on the large scale? What happens if we try to weigh the universe?

If we want to do this, we need to take a closer look at Einstein's mathematics; the equations that describe how the shape of space balances the motion of matter. In principle, each and every solution to these equations is a universe. It may not be the one we live in, but it is a universe nevertheless.

When Einstein formulated his theory, it was generally thought that the universe did not evolve. However, the mathematics pointed in a different

direction, towards a universe that expanded or contracted. In order to fix this problem, Einstein added a term to his equations. The role of this term – today called the cosmological constant – was to make gravity push instead of pull. However, he *was not happy about this change. He could not help feeling that such an ugly thing should not exist in nature.*

A few years later, the astronomer Edwin Hubble discovered that distant galaxies move away from us. The universe is, in fact, expanding. There seemed to be no need for the cosmological constant. It was dropped, but it never quite went away.

If the universe is expanding, it must have been smaller in the past. If you trace this evolution back far enough, do you reach a time when it all began? Was there an initial explosion?

The first evidence for what we now call the Big Bang came when Arno Penzias and Robert Woodrow Wilson found a mysterious noise in their radio antenna. The noise was more intense than expected, evenly spread over the sky, and present both day and night. This cosmic background radiation is a whisper from the Big Bang.

In the last couple of decades, increasingly sensitive space instruments have scanned the cosmos, allowing scientists to probe the universe when it was about 400,000 years old. What they find are tiny variations that have their origin in quantum fluctuations in the very early universe.

We learn that we need three elements to explain the universe.

Imagine making the entire universe into a cake, and slicing this cake into ten pieces of the same size. Only a small part of the cake, half a slice, is made up of normal matter; atoms, molecules, the stars, you and me. A much larger part, two and a half slices, is dark matter, which we don't see, but which we know has to be there from the bending of light from distant galaxies. The rest of the cake, seven whole slices, is dark energy – the cosmological constant – which is needed to explain why the expansion of the universe is speeding up. Something keeps pushing.

The fact that we are utterly clueless about most of the universe, the dark energy, is an enormous problem. A mystery. Something deeply hidden determines how things work. It is not just a test of Einstein's theory of gravity, it is a challenge for the next generation of scientists.

CHAPTER 13

At the speed of thought

IT WAS GETTING hot in the car. Jack rolled up the window and turned the air-conditioning to maximum.

"Where are you taking us?" he asked curiously.

"You'll find out, dad," Liz replied. "Or maybe you can try to figure it out. That would give you something to do."

She gave him a cheeky grin.

Dan was in the backseat, looking out at the desert landscape. He knew where they were going, but he wasn't saying anything. The trip had been his sister's idea. At first he had not been keen on it. It seemed like a lot of hassle, but the more he thought about it the more sense it made. It was the perfect birthday outing for the old man.

After driving through downtown Seattle they had turned east on the interstate. Well over an hour's drive took them across the pine-covered mountains to the edge of the desert. The landscape became increasingly parched as they drove on.

"I have no idea," mumbled Jack, mainly to himself, as he stared at the empty road ahead. "There is nothing out here. Apart from..."

"Oh", he added as it struck him, "the old nuclear reactor site at Hanford and... LIGO?"

"Is that where we're going?" he asked.

At first Liz refused to answer, pretending to focus on the driving. She was annoyed that he had worked it out so easily.

"Yes," she grumbled after a suitably long pause.

"What is this LIGO thing, anyway?" Dan asked from the back of the car.

"Laser Interferometric Gravitational-wave Observatory. Quite a mouthful," Jack replied. "They're listening for gravity signals from the universe.

I'm sure you remember Aunt Helen telling you about the professor's ideas. How space and time stretch and squeeze, like a flexible material. One prediction of the theory is that there are waves of gravity, created whenever heavy objects speed up or slow down. The idea has been controversial. The professor himself couldn't decide if he believed these waves were real. Some people even joked that they move at the speed of thought."

"What on earth is that supposed to mean?" asked Liz.

"I think they were trying to say the waves were figments of the imagination," said Jack. "But they're not. This became clear some ten years after the professor died. People have been building instruments to find them ever since, and just a couple of months ago they did it."

"You mean," Dan asked, "they kept going for almost half a century without seeing anything? Those boffins have some dedication."

"Well, as far as I understand," Jack explained, "these signals are really weak. Space may be flexible, but it is more resilient than rubber, more rigid than steel. It's very hard to bend. The processes that generate the waves may be powerful, but they are rare. And the waves get weaker as they spread through space. So they're tiny when they finally reach us."

*

After nearly four hours in the car, they pulled into the small visitor car park. As soon as they stepped out into the desert heat, they were struck by the scale of the installation. Two fat cement tubes stretched as far as you could see towards the horizon from a central building, forming what looked like a gigantic letter L.

"I don't get it," said Dan, wiping the sweat off his brow. "Why do you need such a big instrument to find something small?"

"I guess we're about to find out, son," said Jack as they walked through the entrance into the cool air of the visitor centre.

"Welcome to LIGO."

The tall man that walked up onto the podium was wearing a colourful T-shirt and a pair of shorts. It may not have looked very smart, but the outfit suited the desert climate perfectly.

The professor would have approved, thought Jack, an amused expression playing on his lips as he recalled the old man in his baggy jumpers and worn-out trousers.

"Before I take you over to the main building, I'm going to explain what it is we do here," the man continued. "What you're about to see is a very unusual telescope. A telescope that explores the dark side of the universe. Exploding stars, colliding black holes, whispers from the massive bang in which the universe was created. That's what we're looking for.

Actually, to be precise, we're not looking. We're listening. Our instrument is more like a microphone than a telescope. We're tracking changes in space and time caused by space-warping events in the distant universe. These gravitational waves are a prediction of Einstein's theory of gravity, where the shape of space determines how stars and planets move.

You may be more familiar with electromagnetism, where a vibrating electric charge produces electromagnetic waves, like light. In a similar way, a vibrating mass leads to gravitational waves. They are created when the earth goes around the sun. As a result, the earth is falling a tiny fraction of a centimeter towards the sun every billion years. It's not a big effect."

The man paused to see if there were any questions.

"That's the challenge for us here at LIGO," he continued. "Catching these tiny waves is far from easy.

Let me tell you about the technology we're using.

The idea is simple. Gravitational waves stretch and squeeze space and time. To find them you need to measure minuscule changes in distance. You need a very precise ruler. We use the light of a powerful laser to measure distances between mirrors. Since the speed of light is constant we can turn the bouncing light into a ruler. We have two rays for comparison, one in each arm of the detector. Those are the long tubes you saw on your way in."

The image on the large screen behind the man changed to show an impressive aerial photo of the installation.

"This technology is actually very old. It was used to measure the speed of light to test the ether theory well over a hundred years ago. It works roughly on the principle that, if two sets of waves are produced and the crest of one is timed to come in the trough of another, the appearance of waves is destroyed. In the case of light, this means we have a dark place. A shadow. This is called interference. We use this interference to check if the distances measured along the two arms are different. If they are, then we've found a gravitational wave.

Our instrument is remarkably sensitive. It is an amazing piece of engineering. It can measure changes in length millions of times smaller than an atom over distances of several kilometers. It's like being able to see the width of a human hair at the distance to our nearest neighbour star through a normal telescope!

But now I think I've talked enough. Let's go over to the control room."

They went back outside and walked across to a large, flat-roofed building. Once inside, the man led the group through to a room full of computer screens.

"We would like to run the instrument pretty much all the time," the man explained. "This is important, because you never know when something will happen in the universe. Or rather, when the signal from something that happened a long time ago will arrive here on earth. Since nothing can travel faster than light, anything we pick up now actually happened in the distant past. In this sense, we are looking back in time."

"Anyway," he continued, "this is the nerve centre of the operation. These screens allow us to monitor that everything is working as it's supposed to. The data is collected and then poured over with a fine-toothed comb to see if we've caught anything."

"We actually catch a lot of things," he said with a wry smile, "but most of them are noise. This is one of the major problems of having such a sensitive instrument. It picks up everything. The tidal pull of the moon. Earthquakes and tremors all over the planet. Even the morning rush hour traffic, the phone ringing, the clouds floating by in the sky and... in this part of the world... tumbleweed.

It's a bit crazy. We need to make sense of this mess somehow. Luckily we have teams of experts working on that. They use some of the biggest computers on the planet to cross check our data against various predictions and ideas. You see, it helps to know what you're looking for. Once you understand what the pattern should be, you can try to match it.

Imagine trying to figure out who said what from a recording of a riotous party, with lots of people talking at the same time and loud music blaring out. You can often make sense of what people are saying because your brain filters out the noise in favour of the voice you're listening to. We're using computers to do that kind of job for us."

"But," asked a man in a striped shirt, "what if your ideas are wrong and you end up looking for the wrong kind of thing?"

"That's a great question," the scientist agreed.

"That's why we're trying to test as many ideas as we can afford to, but we have confidence in our predictions. Astronomers have tracked the effect the energy loss due to gravitational waves has on double-star systems for decades. There are pairs of what we call neutron stars, formed when a normal star dies in a supernova explosion, that gradually spiral together. Just as Einstein's theory predicts. Pairs of black holes should behave the same way. We were hoping to catch the last few seconds of this kind of inspiral and perhaps the final collision, as well."

"And then we did!" With a flourish the man brought up an image on the large screen on the wall. "There it is. The first gravitational-wave signal we've ever caught. Isn't it beautiful?"

"I'm not sure what planet he's from," whispered Dan, just loud enough that Jack could hear him. "It's just a squiggly line."

Before Jack had the chance to respond, the man carried on.

"Yes, I know what you're all thinking. It doesn't look like much. But that single curve tells a remarkable story.

Imagine.

A long time ago, in a galaxy far far away, two black holes edged closer. They danced around each other, drawn together by gravity. In the last few moments the motion grew frantic. A storm of

warped space and time raged as the two objects came together. An energy equal to the obliteration of several suns was released in a fraction of a second. Then it was over. All that remained was a single black hole. And empty space.

The signal moved unchanged over the vast distances of space until, after more than a billion years, it reached the earth. When the signal was created, our insignificant blue planet hosted single cell organisms. When the signal arrived, life had evolved. There was an advanced civilization. A civilization curious about the universe, with the technology to catch the elusive spacetime whisper.

That's quite something, isn't it?"

*

"Phew," Dan sighed when they got back to the car. "That's a lot to take in. Weird stuff."

"You're right," Jack laughed, "there's more to gravity than you think. It does more than just stop you from floating away."

Liz started the car. As they pulled out of the parking lot, Jack turned around to take a final look at the main building.

"You know," he said thoughtfully, "it's been a remarkable journey. I've spent most of my life surrounded by this space-time stuff. I've met so many interesting people. There are so many good stories..."

"Maybe you should write them down," said Liz. "Write a book about it. It could be great."

The suggestion came out of nowhere.

Jack had to think for a while before he decided what he thought about it.

"I think you're right," he smiled. "It could be great."

Author's notes

This story may be a bit unusual.

It started as a vague idea. I thought it would be nice to write something to celebrate the 100th birthday of Einstein's theory of gravity. This might not make sense to most people – you don't just wake up one morning thinking this would be a great plan – but if you have spent nearly three decades working on problems involving the famous professor's brainchild, your mind might be just a little bit warped. Anyway, as soon as I started thinking about it I realised it was going to be tricky.

Eventually, Einstein himself provided the inspiration. He once wrote in the foreword of one of his countless biographies; *"What has perhaps been overlooked is the irrational, the inconsistent, the droll, even the insane, which nature, inexhaustively operative, implants in a individual, seemingly for her own amusement."*

When I started working on this book, after a lot of reading, I had this quote in the back of my mind. I decided to try to tell the story of Einstein and his theory through dialogue and anecdotes, making as much use of actual quotes as possible. The idea was to let the human aspects take centre stage and explain the science as the story developed. This turned out to be a real challenge. Piecing things together, trying to make sense of the science and at the same time not losing the human side, required a lot of hard work. At the same time it was enormous fun.

Basically, the story blends fact and fiction. The science is very much real. So are many of the characters; including Einstein himself (obviously!), his assistant Helen Dukas and the great physicists Robert Oppenheimer and John Wheeler (incidentally, two of my favourites). The narrative is built around things that actually happened, although perhaps not in this particular order or involving precisely these individuals.

Jack is entirely fictional. He had to be...

The Author

Nils Andersson is an expert on Einstein's theory of relativity and related astrophysics. His daily work involves the many extremes of our wonderful universe; black holes, neutron stars and gravitational waves.

He has learned (the hard way) that the devil is often lurking in the detail and that it is important to (try to) figure out exactly how things work. At the same time he likes a good story. Especially the kind of story that explains how things are, even though the story itself may not be entirely true.

Nils has written a series of science-related books for children, based on the somewhat misguided inventor Professor Kompressor. These books are available as e-books, in paperback and the first book is even out as an audiobook so you can listen, laugh and (perhaps) learn.

The illustrator

Oliver Dean is a freelance illustrator and artist from Essex, UK. As part of his diverse practice, Oliver produces live-performance artwork and public participation projects. His clientele are as diverse as his work having collaborated with businesses, charities and educators. He has also authored, illustrated and self-published two picture books for children.

At the time of writing, Oliver is living in Southampton, UK and recent clients include the National Literacy Trust, Boomtown Fair and Hampton Court Palace. He may not able to sing or hold a tune, but he loves to dance, which is also true of his relationship to science and mathematics.

www.oliver-dean.com

Sources

If you want to recreate this effort, you need to trawl through the online archives of the Daily Princetonian, the New York Times, the New Yorker, the Times of London, Scientific American, Time Magazine and others. You need to watch as many TV documentaries as you can find, search the internet for random information and read a number of books. You might want to start with:

Subtle is the Lord & *Einstein lived here*, by Abraham Pais,

Albert Einstein: Creator and rebel, by Banesh Hoffman and Helen Dukas,

Albert Einstein, the human side: New glimpses from his archives, by Helen Dukas and Banesh Hoffman,

Quotable Einstein, by Alice Calaprice,

Einstein as I knew him, by Alan Windsor Richards,

The private Albert Einstein, by Peter A. Bucky,

Inside the centre: The *life of J. Robert Oppenheimer*, by Raymond Monk,

J. Robert Oppenheimer, a life, by Abraham Pais and Robert P. Crease,

Before the day breaks, by Igor Velikovsky,

Geons, black holes and quantum foam: A life in physics, by Kenneth Ford and John Archibald Wheeler,

Magic without magic: John Archibald Wheeler, by John R. Klauder.

Once you are done with the reading, you may be able to identify where the different parts of the story came from.

Good luck!

Thanks

A lot of hard work goes into putting a book together. There is much more to it than a bit of fancy writing. You need to deal with a lot of niggly details to get from that first idea to something that actually works.

This particular book involved a lot of research, piecing parts of the story together, re-writing, re-thinking... It went through a series of more or less radical edits along the way. The story changed, became more personal. Many people provided crucial comments and thoughts, shaping the project in ways I would not have figured out on my own.

I am immensely grateful to everyone that helped. You know who you are!

This has been a long and winding journey, sometimes hard going but always inspirational and great fun. I have learned a lot. Working with Ollie on the illustrations was an absolute highlight. This book would not be the same without his fantastic drawings. I am so very lucky that he put up with my madness.

At the end of the day, this is the best book I could possible write at this point in time. I could keep working on it (probably for ever) but there has to be an end to everything. So this will be the end.

I am really proud of this book and very much I hope you enjoyed it. If you did, and you know someone else that might like it, please tell them about it. You are also welcome to provide any kind of feedback; an online review, a simple rating on goodreads, whatever. Every little thing helps.

Thanks for reading!

N.A. January 2017

P.S. If you have reached this point, there is fair chance that you enjoyed the story. There is, of course, the possibility that you started from the back and this is your first page. If you did, you'll be sorely disappointed because this is not a who-dunnit and there is no suspicious butler to pin the blame on.

Made in the USA
Charleston, SC
20 January 2017